Image Clarity

IMAGE CLARITY

High-Resolution Photography

JOHN B. WILLIAMS

Focal Press

Boston London

Focal Press is an imprint of Butterworth Publishers.

Library of Congress Cataloging-in-Publication Data

Williams, John B., 1941–
  Image clarity : high-resolution photography / John B. Williams.
    p. cm.
  Bibliography: p.
  Includes index.
  ISBN 0–240–80033–8
  1. Images, Photographic. I. Title. II. Title: High-resolution
photography.
TR222.W55 1989
771—dc20   89–1441
                                                  CIP

British Library Cataloguing in Publication Data

Williams, John B., *1941–*
  Image clarity: high-resolution photography.
  1. Photographic images. Quality
  I. Title
  771'.53

  ISBN 0–240–80033–8

Butterworth Publishers
80 Montvale Avenue
Stoneham, MA 02180

10 9 8 7 6 5 4 3 2 1

Printed in the United States of America

hīgh-res·ō·lū′tīon pho·tog′ra·phy, *n.*

1. the art or practice of recording photographic images accurately and with detail

2. an approach to the craft of photography having as its goal the reproduction of fine subject features with exquisite sharpness

3. photography that achieves consistent image clarity and that results from maintaining high standards of technical excellence in the application of photographic skills

Contents

Preface

The capacity of a photograph to capture the details of nature can be as compelling as its capacity to capture nature's beauty. Many photographers have been drawn to the quest—not in search of detail for its own sake, or the search would have ended with large-format cameras. Instead, they are drawn to reach the limits of the photographic process—to get from it all the detail it can deliver. This is what high-resolution photography is about.

Traditional photographic literature would have us believe that anyone can create photographs of stunning detail and sharpness simply by switching to a fine-grain film, using a sharp lens, and mounting the camera to a tripod. But it is not so. Even with a near-perfect imaging system, what seems to be a law of natural perversity makes it all but impossible for fledgling photographers to exploit the system's full potential. They find that the most minute kinds of image degradation can be recorded in their photographs. The great clarity of reproduction of a superior optical system only makes tiny image defects more easily visible. The better the imaging system, the more challenging it is to exploit its capabilities. Clearly skill and technique are as decisive as equipment.

In seeking the greatest possible sharpness and detail in their photographs, aspiring high-resolution photographers ultimately discover that the techniques offered in traditional photographic books do not always succeed. Several factors help to explain why.

- Many writers and accomplished photographers have little concern about pure technical quality. The ability to make sharp, clear photographs, although a basic skill, is neglected and underrated and gets far less attention than it should. Even worse, some writers depreciate it by exhorting image softness and ambiguity.
- Those who are concerned with technical quality disagree about proper technique. This disagreement arises from differing stylistic

tastes, from the desire for efficiency and economy in business, from the misdirection caused by overly ambitious advertising claims of product manufacturers, and from simple disregard for facts.

- Traditionally photographers have been taught to focus on the kinds of degradation that operate within the macrostructure of the image—graininess, focusing errors, image motion, and the like. The techniques offered to counter this kind of degradation are certainly basic to high-resolution photography, but these techniques are not enough when one must go to the limits of resolution of the silver-halide process or must perform predictably under impossible circumstances.

- Images are degraded on a microscopic level in ways that are generally dismissed. The microstructural damage—done by the degradation of contrast, by diffraction, and by ground vibration—is so subtle as to be inconsequential. Left alone, however, such damage accumulates. The more perfect the imaging system, the more frustrating this subtle degradation can be. In high-resolution photography, degradation of all kinds must be controlled, and all must be controlled concurrently.

- Certain physical principles apply to photography (and indeed to image formation in general) that explain why photographs are degraded during the imaging process. Imaging techniques that ignore these principles invite degradation and make it inevitable.

The Need for This Book

This book evolved out of research and experiments undertaken (1) to consolidate current knowledge about clarity in photographic processes, (2) to gain insight into the principles and theory of photographic sharpness and resolution, and (3) to sort out from the conflicting photographic philosophies techniques that are consistent with these principles.

In many applications of photography, the accurate rendition of visual information with full resolution of detail is the only goal. Scientists engaged in field studies—archaeologists, geologists, zoologists, and the like—rely heavily on photographs as visual records of their studies. See Plates 14 and 15.* Also criminal investigators, accident investigators, forensics specialists, legal photographers, military-surveillance photographers, and other wizards who push photography to the limits of its resolution potential must capture the details of their subjects accurately, sometimes under challenging conditions. Their increasing reliance on 35 millimeter cameras and the substantial enlargement of their negatives make high-resolution techniques essential to the success of their work.

The resolution potential of still photography is becoming as im-

*Note that all plates are in an insert in the middle of the book.

portant to business and industry as it is to the sciences, not only as a medium for publicity and promotion, but as an investigative and documentary tool for training, testing, and control. It is a detection device used for quality control in production and a blessed substitute for vision in situations where dangerous, transient, or fast-moving events must be recorded.

General photographers as well, from novices to practicing professionals, will find the principles and techniques of high-resolution photography useful for strengthening their basic skills. After all, mastery of high-resolution techniques and an understanding of the principles of accurate image formation should be the foundation on which all training in the craft of photography is built.

Emphasis herein is therefore on achieving image clarity and detail. The underlying assumption is that, above all else, the reader wants the sharpest images possible. Not that high-resolution photography is the solution to every photographic assignment. Indeed, total reliance on high-resolution techniques by certain commercial and portrait photographers would be a mistake. Photography is used as both a technical and expressive medium in which the balance between artistic and technical considerations can change from one assignment to the next. Nevertheless, pure technical quality is the objective in this book.

In a photograph of a criminal suspect used as evidence in a judicial proceeding, in a close-up study made to record the anatomy of a cicada, or in a photograph of a distant galaxy made to fix the location of a super nova, the success of the photograph will not be decided on artistic merit. Granted, a well-composed, visually appealing rendition will not detract from the technical accuracy of the photograph, but the purpose of the image in these instances can be met without considering artistic principles. There are occasions when the requirement that an image be clear alone determines the direction a photo session will take. These are the occasions addressed herein.

The resolution of realistic detail is, after all, the great strength of photography. Photography records fine details and isolates moments in time and space more completely and more accurately than any other graphic process.

As it turns out, there are advantages to using the approach adopted herein:

- A single, unifying theme simplifies the discussion making it easier to follow.
- The narrow scope makes possible a thorough, exhaustive discussion of the processes associated with accurate image formation. As a bonus, contradictions are more easily resolved and confusion is more easily eliminated.
- From this consolidation of information, new relationships are seen more clearly.

There has long been a need for such an approach. The proliferation of photographic styles, divergent philosophies, and opposing points of view seen among respected, accomplished photographers, though understandable and perhaps inevitable, can be perplexing to novice photographers seeking a model on which to base their style and seeking to understand the merits of various techniques. For the novice's sake, a focused approach to the craft of high-resolution photography is overdue.

Also, scant literature that gives serious attention to high-resolution techniques is available. What is available falls at one of two extremes. Literature directed at the beginning photographer often simplifies the discussion to the point of triviality. Literature directed at photographic scientists and engineers is more accurate and complete, but the reader needs the training of an engineer to follow what is being said. Little else has been available for the informed practicing photographer.

Acknowledgments

I would like to thank Bill Becker and Mike Fletcher for their editorial assistance and Susie Shulman and the staff at Focal Press and Jane Richardson and the staff at Bywater Production Services for their help.

Chapter 1

Introduction

What Is High-Resolution Photography?

In a literal sense high-resolution photography is about the photographic reproduction of a subject in full detail, down to its smallest features. In a more general sense it is about image accuracy, clarity, and sharpness, and about the faithfulness of an image to its subject. It is an approach to the craft of photography having as its goal the reproduction of fine subject features with exquisite clarity and sharpness. It is a strategy intent on achieving technical excellence by applying the highest standards of photographic craftsmanship.

How This Book Is Organized

The text is divided into three parts: Theory of High-Resolution Photography, Mechanisms of Clarity and Degradation, and High-Resolution Techniques. This kind of organization separates the theoretical from the practical discussion, allowing nontechnical readers to go directly to the chapters that interest them most.

Part I explores the origins of the physical and mathematical principles that govern image formation and it examines the implications of these principles. Part II examines in depth the processes that can degrade an image, showing how they operate within the microstructure of an image to destroy detail and diminish sharpness. Part III is a practical compendium of high-resolution techniques.

Exciting concepts are explored in Part I. Image clarity and visual quality are defined in Chapter 2 so that the reader begins with a full understanding of what makes an image appear sharp and clear. The reader learns more about the concept in Chapter 3 where image clarity is discussed in conjunction with four important image-evaluation techniques: resolving power, acutance, modulation-transfer functions, and information theory.

The converse of image clarity, image degradation, is treated in Chapter 4. Image degradation is a major theme in this book. The image-degradation formula, introduced in Chapter 4 and elaborated in Chapter 5, is a mathematical expression that describes the cumulative effect that isolated sources of degradation have on an image; this formula clarifies certain aspects of image formation and is the basis for much of what follows herein.

Contrast, another major theme, is discussed as early as Chapter 3 in relation to its influence on acutance and the modulation-transfer function. This theme is returned to in Chapter 7 wherein a study of the microstructure of the photographic image reveals the mechanism by which the deterioration of image contrast destroys image clarity and lowers the resolution of fine detail.

The practical limitations of human vision are examined in Chapter 6 in an effort to determine how good an image must be to seem sharp and clear and how bad an image must be to seem badly blurred. The theoretical discussion of Part I ends in Chapter 8 where the reader learns to measure photographic quality in terms of visual clarity, picture quality, or legibility. Therein one will find meaningful ways to express numerically the magnitude of image clarity and image blurring.

Part II contains detailed explanations of the problems encountered by and the tools available to high-resolution photographers. The need for this division and the reason for the book's emphasis on image degradation is justified and is fully demonstrated early in Part I. Each subject in Part II is examined in enough detail to demonstrate how it can alter the microstructure of an image to lower image clarity or to enhance it. Those properties of light that influence clarity, such as quality and direction, are discussed in Chapter 9. The contrast theme reappears in Chapter 10 where it is presented from two perspectives: that of its effect on tone control and gradation, and that of its capacity to reduce resolving power. Chapters 11 through 14 deal with emulsions, the chemistry of development, color, lenses, and the modulation-transfer function. Part II ends with Chapter 15, a look at image motion.

Part III presents practical, tested, high-resolution techniques. Although Part III has the structure of a traditional photographic handbook, more so than Parts I and II, it is still unique in using the principles established in Part I to address the problems examined in Part II. Beginning in Chapter 16, many useful techniques are revealed. Chapter 17, for example, in conjunction with Chapter 12, provides a thorough introduction to the chemistry of development, enough to allow readers to customize high-resolution developer formulas. The remaining chapters deal with routine tasks associated with high-resolution imagery: camera skills, tripod use, lighting, and so forth.

Part I

Theory of High-Resolution Photography

Several physical principles govern the process of image formation. These principles impose limitations on the accuracy and clarity of images, whether they be formed in a camera, on canvas, on a video screen, or in the eye. Understanding what these principles are and how they operate is the first step toward getting the best image clarity that photography can offer.

Chapter 2

Visual Quality
Understanding Resolution and Sharpness

To master the techniques of image clarity, one must eliminate image softness and blurring, the enemies of visual quality, but one must first understand what these things are and how they result. What determines visual quality? Although anyone with normal vision can distinguish a clear image from a blurred one, many people would find it difficult to explain the difference. What is it that allows a person to perceive that one photograph is sharp and clear and another soft and blurred? Before you set out to improve your photographs, perhaps you should take time to determine precisely what needs to be done to them.

Image Clarity

High-resolution photography is essentially about image *clarity,* a very simple concept. An image is clear when its details match what is seen at the original scene. Clarity is related both to the fineness of detail, indicated by image *resolution,* and to the visibility of image features. *Visibility* is determined by how much a feature contrasts with its surroundings.

Sharpness is an aspect of image quality quite different from resolution. Whereas resolution is determined by how small image features are, sharpness is determined by how distinct their outlines are. Resolution is related to the number of lines in an image; sharpness is related to the quality of those lines. The perception of sharpness is influenced mainly by edge contrast, sometimes called edge *definition*. Edges are sharp when they do not spread into or merge with the background or with adjacent tones. A sharp edge has a clean outline at which the transition between tones is abrupt. See Figure 2.1.

Hardly interchangeable concepts, sharpness and resolution are determined by entirely different image parameters. Indeed, it is possible to photograph a subject once to produce a soft image with much detail and again to produce a sharp image with less detail. (See Plate 1.) More often though, whenever image quality declines, it declines in both sharpness and resolution. Most of the imaging errors that reduce sharpness have an equally harmful effect on resolution.

Subjective Evaluations of Clarity

The ability to perceive clarity, to detect sharp edges and fine detail, is a fundamental property of vision. It is not a learned skill so much as a natural and inherent feature of perception. An observer can visually judge image quality based on how clear an image seems to be. In doing so, he or she mentally compares the image with others seen in the past and ranks it higher or lower in clarity. Visual assessment is simply an intuitive determination of preference for one image over another. It is a convenient, natural way to evaluate not only images, but films, lenses, and other photographic components. To evaluate equipment, one simply takes a photograph and visually compares it to those made with another model or brand of the item used.

As simple as it is in concept, visual examination is by far the most

FIGURE 2.1
Edge sharpness

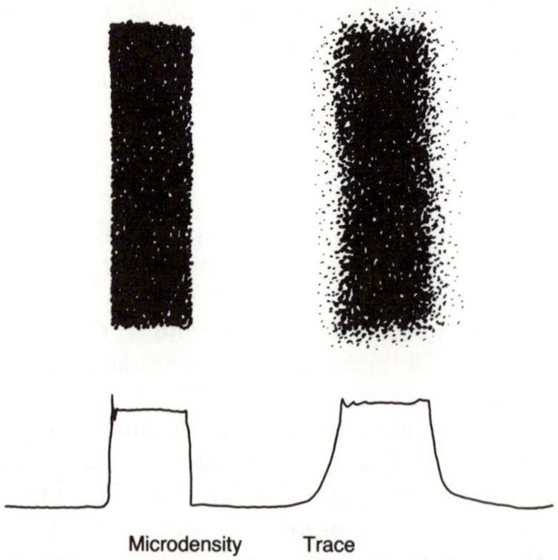

Microdensity   Trace

The difference between a sharp and a blurred image shows up in the accuracy of the reproduction of edges. An image is sharp when the change in edge density is abrupt (top left) instead of gradual (top right). The density trace across a sharp edge is nearly vertical at the transition between tones. The density trace across a soft edge is rounded.

important method available for determining picture quality or, by extension, for rating the systems and components used to make photographs. Despite today's sophisticated optical measuring devices, mechanical measurements of quality can never completely substitute for human judgment and visual perception.

Most observers are quite reliable in their visual differentiation of images that are sharp and clear and those that are not. Visual assessments are usually both consistent and useful, but the accuracy of an assessment is high only when there is a pronounced difference in quality between the images compared. Inconsistency in the ability to see small variations in quality shows up in two ways. First, different observers may disagree about how an image should rank among others of nearly equal quality. Second, the same observer may rank an image differently on different occasions. The outcome of visual comparisons at this level of difference may thus be unrepeatable. Consequently visual assessment, lacking in reliability and predictability, is hardly a scientific method of evaluating images.

Psychological factors also influence the perception of image sharpness and may lead one to rate an image higher or lower than indicated by objective measures. Again, sharpness is perceived from specific clues in an image, the most important being high edge contrast. Edge contrast is so essential to this perception that when edges are missing in an image, sharpness cannot be evaluated at all. A photograph of a blank sky appears neither sharp nor blurred because there are no clues in a blank sky by which to judge edge quality. Furthermore, if objects are present, all their edges may be too soft, as with a cloud or a wisp of smoke. One cannot determine, just from a photograph, whether the rendition of such a subject is accurate or poor. The absence of sharpness clues or the presence of ambiguous clues can predispose one to rate the quality of a system low even though the image it produces closely matches the subject.

For detail and sharpness to appear in a photograph, it is not enough that the camera used be of high quality; the subject must contain features with sharp edges, fine detail, and rich textures. If the subject is soft and featureless, images will be soft and featureless and may incorrectly appear to have been photographed with an inferior system. Unless the subject provides a tough test of reproduction accuracy, visual indications of quality can be misleading. Visual assessments based on inaccurate impressions of sharpness can result in faulty comparisons that may be at odds with ratings derived from objective measurements.

Because a photographer cannot always personally test each lens and film of interest, assessments made by others are often used as a basis for choosing one component over another. It is best that such ratings be both *objective* and *numerical*. Objective ratings are important because they can be confirmed by independent evaluators if necessary. If several unbiased photographers rate a system simi-

Objective, Numerical Evaluations of Clarity

larly, the likelihood is reduced that initial results were influenced by subjective and psychological impressions. One can be more confident that the system will be rated much the same during succeeding evaluations.

Numerical ratings indicate not only that two systems differ in quality, but by how much. More importantly they can reveal mathematical relationships. By replacing inexact verbal labels with precise measurements, numerical ratings let photographic scientists and engineers apply higher-order mathematics to the design and construction of photographic components. Numerical methods have, at the same time, given practicing photographers greater insight into the true nature of image clarity and have offered alternative ways to study it.

Chapter 3

Analyzing Visual Quality
The Mathematics of Image Clarity

A good way to learn about image quality is to study how it is measured. Such an approach will also acquaint you with the mathematics of image clarity.

Over the years, several methods have been advanced for measuring and numerically rating what is perceived as visual quality. A goal of each has been to replace verbal, subjective descriptions, such as *clarity, definition,* and *sharpness,* with objective ratings that correlate with human visual perceptions. *Resolving power* and *acutance* are the result of early efforts. Resolving power correlates with the fineness of detail; acutance correlates with sharpness. Both concepts provide valuable insight into the nature of image quality. They have limitations, however, and other methods, like modulation-transfer functions and information theory, have recently come to the fore. These techniques, the old and the new, are explored in detail in this chapter.

Resolving power is one of the earliest methods devised for numerically rating optical systems and evaluating image quality. Resolving power measures the capacity of a system to reproduce closely spaced lines as separate and distinct line images. It is usually stated as the number of lines per millimeter a system resolves. When a pattern of lines and spaces is reduced in size, the pattern eventually becomes so small that an optical system cannot reproduce each line separately, but instead reproduces the pattern as a uniform gray tone. Resolving power is determined by the line spacing just barely resolvable before this merging of lines and spaces occurs.

Resolving Power

The test patterns or targets used in resolving-power tests consist of alternating bars and spaces. See Figure 3.1. Within individual target groups, each space is as wide as a bar. Resolving power, when determined from such a target, is given by the relationship:

$$R = \frac{1}{d}$$

where R is resolving power in lines per millimeter and d is the distance in millimeters separating bars that are just resolved. The separation between target bars is measured from the center of one bar to the center of the next. See Figure 3.2.

Resolving power is a useful procedure that accurately measures the capacity of a system to record fine detail. Systems with high resolving-power ratings are well suited to recording subjects rich in texture and detail. Also, because resolving-power ratings are numerical, they allow direct comparisons between systems. The enduring value of resolving power lies in the simplicity of performing a resolving-power test. It requires no special test equipment other than a standard resolving-power test chart.

Resolving power is versatile. It can be used to measure the quality of a lens or a film separately or that of an entire photographic system. To test a lens, the optical image of the test chart is examined

FIGURE 3.1
Resolving-power test chart (target design)

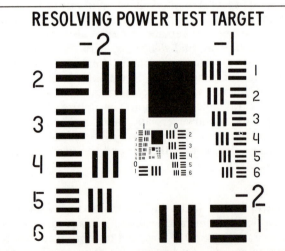

Resolving-power test charts are available in various sizes and designs to test for the presence of specific image errors. A practical test chart can be nothing more than the classified section of a newspaper, although such a chart will help to determine only that one component is better than another. Charts having bar-space targets are used when resolving power is to be measured in lines per millimeter. (From *Photographic Materials and Processes*, Stroebel, Compton, Current, and Zakia, courtesy of the authors.)

using the lens to be tested. Resolving power is then found from tables or formulas that take into account the focal length of the lens, the distance from the lens to the test chart, and the size of the smallest target pattern resolved.

The system's resolving power is found from a photograph of the test chart made at a specified distance. The developed film, or an enlarged print, is examined with a magnifier to find the smallest pattern in which lines and spaces are clearly visible. When resolving power is determined from a print, the image will, of course, have been affected by each component of the photographic system: film, lens, developer, enlarger, and so forth. The rating is then said to represent the system's composite resolving power.

Finding the resolving power of a film alone is a more complex procedure requiring both the lens and system tests, described above, to be conducted first. From the resolving power of the lens and the system, the film's resolving power is estimated using the empirical relationship:

$$\frac{1}{R^2} = \frac{1}{r_o^2} + \frac{1}{r_e^2}$$

where

R = the resolving power of the system,
r_o = the resolving power of the lens, and
r_e = the resolving power of the film.

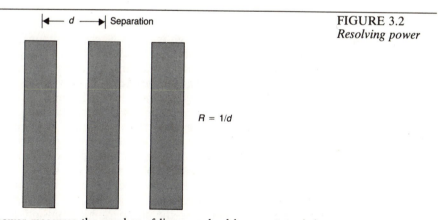

FIGURE 3.2
Resolving power

Resolving power measures the number of lines resolved by a system; it is computed either by counting the lines or by measuring the separation between lines and then calculating the reciprocal of line separation. Notice that line separation is not the width of a space but the distance from the center of one line to the center of the next. Because line separation equals the width of a line plus a space, resolving power is often referred to as "line-space pairs" or "line pairs" resolved.

Resolving power is an important tool for photographers. It has been in use longer than other numerical image-evaluation procedures. It is well known and its methods are widely publicized. It culminates in numerical ratings that allow direct, quantitative comparisons between systems and that correlate well with visual impressions of image detail. It can be used to evaluate individual components separately or to evaluate entire photographic systems.

Nevertheless, resolving power is an imperfect measure of image quality. The outcome of resolving-power tests has been found to vary from one independent evaluator to another, even where both were known to be impartial. Variations arise because the evaluator must determine visually which target group has been resolved. Final selection is a matter of judgment and depends on the experience and visual acuity of the evaluator. The outcome can also be affected by vibrations, lighting conditions, the contrast of the test target, variations in spacing between target bars, or variations in the layout of the test chart. The outcome of tests conducted on films is further affected by variations in exposure, the degree of development, the constitution of the developer solution, and the precision of focus.

The important limitation of resolving power is that it does not measure fully the characteristics that influence image clarity: it does not measure the capacity of a system to produce sharp images. It sometimes happens that images produced by low-resolution systems are actually sharper than those made with high-resolution systems. The resolving power of a system correlates well with its ability to reproduce fine detail but does not always indicate the system's ability to produce good edge contrast.

Acutance

As noted earlier, an image is sharp where the boundaries between adjacent features and objects are distinct and clearly demarcated. Sharpness decreases when tonal areas blend into one another or when density at edges changes gradually by way of successive shades of gray. The softer the edge, the wider the transition zone between adjacent tones. Acutance describes this density change mathematically and provides a numerical rating designed to correlate with the quality perceived as sharpness. Although numerical acutance ratings are rarely used today and are not included on film data sheets, reviewing the method used to obtain them will greatly improve your understanding of the nature of edge sharpness and image clarity.

Conveniently, the acutance of a film can be computed using measurements taken from the photographic emulsion independently of the optical system. To measure the acutance of a film, a sharp knife edge is contact printed onto the film using either collimated light or a point-source lamp. See Figure 3.3. After the film has been processed, the density change across the edge formed by the knife blade is measured using a microdensitometer and plotted on a graph. Before being plotted, the density readings are averaged to eliminate the fluctuations caused by emulsion grain and to smooth the plot line. Density is plotted along the vertical axis; the distance from the edge is plotted along the horizontal axis. The graph so obtained is called the *gradient curve* of the emulsion. See Figure 3.4.

The gradient curve associated with acutance may seem similar in appearance to the characteristics curve of an emulsion (see Figure 9.5), but the two should not be confused. The characteristics curve shows how density changes with changes in exposure level. The density variations in acutance gradient curves are caused not by exposure differences, but by the spreading of light within the emulsion. This spreading occurs when light is reflected off emulsion crystals and is sent into areas shaded by the knife blade, areas outside the original path of the exposing light.

Acutance can be computed from the gradient curve using an empirical formula. Before the formula is applied, the horizontal axis below the density graph is divided into an arbitrary number of equal segments N, and the incremental change in density ΔD on the gradient curve associated with each segment ΔX_i is measured. See

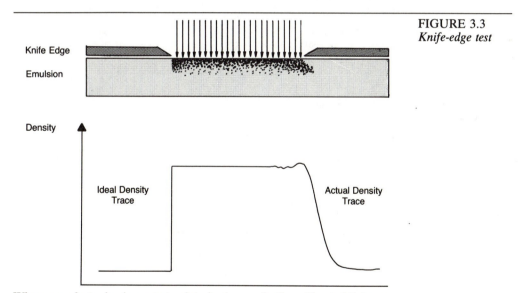

FIGURE 3.3
Knife-edge test

When a perfect edge is contact printed using perfectly collimated light, one might expect it to be imaged with perfect sharpness. In reality, image density at a photographic edge does not change abruptly from dense to clear (shown in the ideal density trace on the left) but trails off gradually (shown in the density trace to the right). This rounding of the density trace occurs because emulsion crystals below the knife edge, crystals that are outside the direct path of the exposing light, are exposed by light reflected from neighboring crystals.

Figure 3.5. The Greek symbol Delta, Δ, signifies "the incremental change in" the variable that follows it. Acutance is found by applying the formula:

$$A = \frac{1}{N} \sum_{i=1}^{N} \left(\frac{\Delta D_i}{\Delta X_i}\right)^2 (D_{max} - D_{min})$$

where

N = the number of segments that divide the horizontal axis,
ΔX = the width of each segment on the horizontal axis,
ΔD = the change in density associated with a given ΔX,
D_{max} = the maximum density value, and
D_{min} = the minimum density value.

In simple terms, acutance can be considered the product of edge gradient and edge contrast. The factor in the equation represented by $\Delta D_i / \Delta X_i$ measures the average change in density at a particular distance from the edge and gives the *slope* or *gradient* of the curve at that location. The greater the acutance, the steeper the slope and the greater the computed value becomes.

The last factor in the formula, $(D_{max} - D_{min})$, measures the difference in density between the high- and low-exposure areas. The greater the density difference between two tones, or the higher the

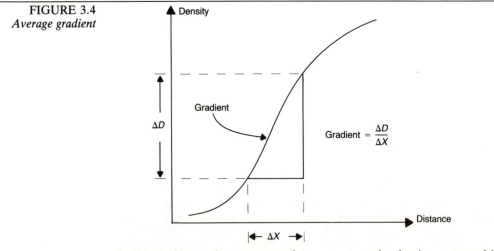

FIGURE 3.4
Average gradient

Gradient refers to the steepness of a segment on the density curve and is measured by dividing a vertical rise on the curve ΔD by the corresponding horizontal change in distance ΔX. Gradient shows how fast density increases or decreases at a given distance from the edge.

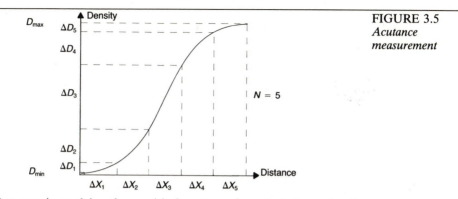

FIGURE 3.5
*Acutance
measurement*

The first step in applying the empirical acutance formula is to mark off equal segments on the horizontal axis below the gradient curve and compute the gradient associated with each segment. These gradient values are squared and averaged, and the average is multiplied by the density range of the emulsion to yield an acutance rating.

contrast, the greater this factor becomes, and again the value computed for acutance increases.

Acutance, like resolving power, gives a meaningful indication of image quality; it correlates well with the way the eye perceives sharpness and it provides a numerical rating. Unlike resolving power, it is not influenced at all by subjective impressions, since it is derived from instrument readings that require no visual judgment on the part of the evaluator. Acutance is a numerical and objective rating.

Acutance is not easily measured, however, making it difficult to use in routine evaluations of image quality. It requires laboratory measurements involving the use of an expensive microdensitometer. Typical photographers may not own such an instrument; few will find it beneficial to acquire one just for this purpose. Acutance also shares a weakness with resolving-power ratings: it does not completely account for image clarity. Though it correlates closely with sharpness, it does not indicate how good resolving power is or how accurately a system reproduces fine detail. Thus acutance cannot stand alone as a measure of image quality.

Modulation-Transfer Functions

The use of the modulation-transfer function (MTF) in evaluating photographic quality is growing in popularity. The intriguing curves associated with the MTF now appear frequently in the promotional literature of lenses and films. See Figure 3.6. As will soon be seen, modulation-transfer functions overcome a weakness of resolving power and acutance by providing a more complete indication of image quality. Also, being generated mechanically, they are objective.

The concept of modulation transfer was first used to solve problems in electrical engineering. Consequently, much of the language associated with this concept—that of signals and noise, frequency and amplitude—has been borrowed from the field of electronics. Some of the more common terms, seen more frequently in photographic literature and used in this and in later chapters, are briefly introduced here.

A *signal* is a means for conveying information. It functions in either an input or an output mode. An example of an electronic input signal is a radio wave that arrives at a radio antenna; its corresponding output signal is the sound wave emitted by the radio. In photography an input signal is the image that originates at the subject or the light that enters a lens to form an optical image. The output signal is either the real or virtual image.

The basic properties of a signal are its frequency and its amplitude. See Figure 3.7. The *amplitude* of a radio signal is the difference between its maximum and minimum signal strength and is an indication of the power of the signal. The amplitude of a photographic signal is related to its luminance range or density scale. A weak signal is one with low image contrast.

Frequency is a measure of the number of repetitions of the signal over a given linear distance (spatial frequency) or over a given period of time (temporal frequency). A signal's wavelength is sometimes substituted for its frequency. *Wavelength,* the reciprocal of frequency, is related to the size of a repeating element. The parameters in image formation that correspond to frequency and wavelength are resolving power and detail separation. If the input signal arriving at a photographic system is the image of a standard resolving-power bar-space test target, the spatial frequency of the input

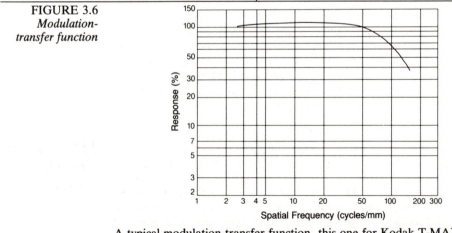

FIGURE 3.6
*Modulation-
transfer function*

A typical modulation-transfer function, this one for Kodak T-MAX Professional Film 5052. (Reprinted courtesy of Eastman Kodak Company.)

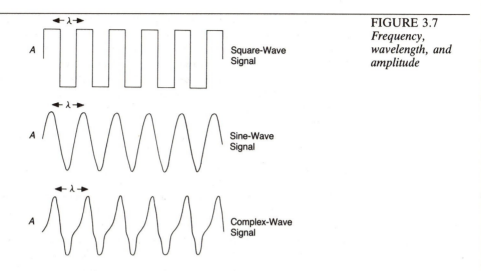

FIGURE 3.7
*Frequency,
wavelength, and
amplitude*

A   Square-Wave
Signal

A   Sine-Wave
Signal

A   Complex-Wave
Signal

Frequency f and wavelength λ are reciprocal properties of a signal: $f = 1/\lambda$ and $\lambda = 1/f$. Wavelength measures the distance from one signal peak to the next along the line of propagation. Frequency measures the number of repetitions of a signal in a given period of time (temporal frequency) or over a given linear distance (spatial frequency). Amplitude, A, refers to the strength of a signal as measured from its minimum power level to its peak.

and output signals will correspond to the number of line-space pairs in the target, and wavelength will correspond to the separation between lines.

Noise, which interferes with a signal's purity and clarity, consists of the unwanted changes imposed on the signal by spurious secondary signals or by imperfections in a transmitting or receiving system. The noise in a photographic system arises primarily from the granular structure of the film and from imperfections in the imaging system.

Modulation refers to such fluctuations in frequency, phase, or amplitude as are necessary to encode information onto a signal. The related photographic concept is that of image formation itself.

Much can be learned about modulation-transfer functions by examining how they are generated, so let's do that now.

The test targets used to generate the MTF differ in design from resolving-power test targets in two ways, both of which are important to the outcome. First, they are reproduced on a transparent base so that, by illuminating them from behind, target brightness can be kept constant over the entire frequency range tested. Second, the bar-space patterns used in resolving-power tests are replaced by sinusoidal patterns in which the target patterns change gradually in

Generating Modulation-Transfer Functions

density from clear to dark in proportion to a sine function. (See Figure 3.10.) These sinusoidal patterns vary in wavelength; pattern elements and the spacing between them are quite large at one end and become progressively smaller along the length of the test target.

The significance of the sinusoidal design of the MTF test target arises from the somewhat sinusoidal manner in which photographic images are degraded. Consider what happens in a photograph of bar-space patterns such as those on a standard resolving-power test chart. A density trace taken across the image of these targets inevitably yields a sinusoidal-like plot line, even though the original input signal begins as a square wave. See Figure 3.8. If square-wave signals could be recorded perfectly on film, the density trace of the image would show an instant change or a vertical rise or drop in density at each edge. The shape of the trace would be rectangular or square—hence the name square wave. Density traces taken across such edges in actual photographs never show instantaneous density changes. Real traces have sloping or curving sides instead of vertical ones and rounded instead of square corners. See Figure 3.9.

The density traces associated with low-frequency square-wave signals or large bar patterns are much less affected by this kind of degeneration than are the density traces associated with high-frequency signals. At low-input frequencies and in systems wherein image degradation is small, the output signal is slightly rounded at the edge but retains the basic shape of a square wave. At higher frequencies or as degradation increases, the corners of the output

FIGURE 3.8
*Degeneration of
bar targets*

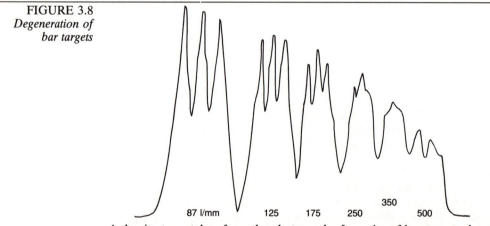

A density trace taken from the photograph of a series of bar targets shows how image accuracy degenerates with an increase in the spatial frequency of image detail. A significant feature of this degeneration is the loss of contrast, first among individual bars, then at higher frequencies among groups of bars.

signals are decidedly round and the sides are sloped instead of vertical.

As degradation increases further, the image of a bar target degenerates entirely into a sine wave. The corners are then completely rounded and the sides of the curve are decidedly sloped. With further increases in the frequency of the signal or with further degradation in signal quality, the output signal retains its basic frequency as well as its sinusoidal shape, and the difference in the quality of transmission shows up as a change in the signal's output amplitude.

Image degradation alters a square-wave signal by first changing its shape from square to sinusoidal, then by reducing its amplitude.

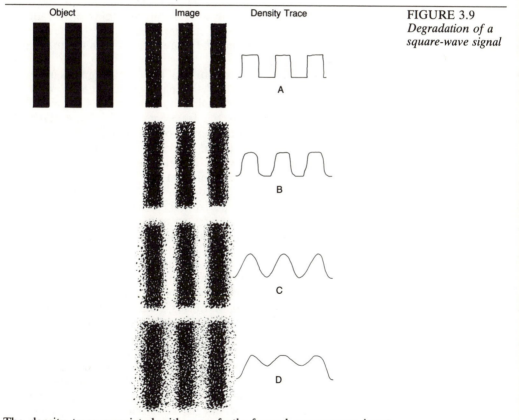

Object   Image   Density Trace

FIGURE 3.9
Degradation of a square-wave signal

The density trace associated with a perfectly formed square-wave input signal would have a vertical gradient and square corners (A). Rounding at the corners indicates that signal quality has declined (B). On being degraded significantly, the output waveform degenerates into a sine wave in which all corners are rounded and the gradient is sloped all along the curve (C). Further degradation in the system shows up as a reduction in the amplitude of the sinusoidal wave (D).

FIGURE 3.10
*Sinusoidal test
target*

The bars of a sinusoidal test target change gradually in density in proportion to the trigonometric sine function. A typical target has three bars of each frequency. (Reprinted courtesy of Eastman Kodak Company.)

The test target used to generate the MTF is sinusoidal at the start. See Figure 3.10. Thus errors in the imaging system that degrade the output signal do so by reducing its amplitude alone. Reductions in signal strength thereby correspond more directly with the magnitude of image degradation.

It is the reduction in the amplitude of the output signal, equivalent to the reduction in contrast of the photographic image, that the MTF shows. It measures the decline in the transfer of contrast from the subject to the image as the frequency of the input signal increases. It shows how well the power of the original signal is transferred by an imaging system and retained in the output signal and thus indicates how accurately the fine and coarse features of a subject are reproduced. At any given point on the MTF, a vertical coordinate indicates the contrast retained in the image at the frequency shown by the corresponding horizontal coordinate. Contrast is often expressed as a percent and is arrived at by computing the ratio of image modulation to object modulation, M_i/M_o. In a perfect imaging system the contrast of the input target would be transferred to the film at 100 percent of its original value; image modulation would be equal to object modulation. A smaller transfer indicates that the signal has weakened and that image quality has degenerated.

When the MTF of a lens or film is generated, the brightness of the test target is held constant at all target frequencies to provide a constant input signal strength. This is accomplished by illuminating the transparent target from behind using a light source of constant intensity. As a result of this precaution the value of M_o, the denominator in the contrast ratio, does not change during the test so that the ratio M_i/M_o at any frequency will always be directly proportional to M_i, the brightness of the image. The value of M_i/M_o, and consequently of the modulation-transfer function, is therefore obtained directly by reading image brightness; no other computations or measurements are necessary. The nature of the process suits it to automated measurement, and, indeed, instruments are available that generate the MTF directly from microdensitometer readings of a film, and optical-transfer functions directly from an optical image.

In the modulation-transfer functions of lenses, contrast at zero frequency is normalized to 100 percent. The modulation-transfer function of a lens will therefore not exceed 100 percent. The modulation-transfer function of a film may or may not exceed 100 percent at zero frequency depending on how the film is developed.

The contrast of a lens at low-input frequencies, which indicates how well it reproduces dominant shapes and large features, is always relatively high. Most lenses, even simple ones, are capable of reproducing major image features accurately.

High-frequency contrast, on the other hand, indicates how accurately a lens mediates fine detail. As the frequency of an input signal increases, the contrast of the output signal invariably decreases. The better the lens, the greater its contrast at high frequencies and the greater its resolving power. In all lenses, however, contrast declines as the input frequency increases and drops toward zero as the resolution limit of the lens is approached.

The greater the modulation-transfer function, the more accurately the lens will pass the image. An MTF of 100 percent is ideal, but the imaging quality of a lens will be comparatively good whenever its MTF is greater than 70 or 80 percent. As the MTF contrast of a lens declines to about 30 percent, its imaging clarity will range from fair to marginal. Yet although imaging quality may be poor at 30 percent contrast, image features reproduced at this level may still contain usable information and discernible detail. At image contrast of 20 percent or less, however, high-frequency detail will in all likelihood cease to be visually discernible.

Modulation-transfer functions also reveal something about a system's sharpness. They do so by showing the range of frequencies passed by the system. It is theoretically possible to synthesize a regular square wave, the equivalent of a perfectly sharp edge, using only sine waves. See Figure 3.11. To do so, a basic sine wave is selected whose frequency equals the frequency of the square wave to be synthesized. To this basic wave are added sine waves of increasingly smaller amplitude in each of the odd harmonic frequencies of the basic wave. The properties of the harmonic waves needed for this synthesis are such that, as the frequency of a wave increases, its amplitude decreases. With the addition of each harmonic wave, the slope of the composite wave gets steeper, the peak gets flatter, and the ripples in the peak get smaller. To synthesize a perfect square wave, however, an infinite number of harmonics must be used. By extension, a perfectly sharp imaging system must, in effect, have an infinite frequency response.

As a practical matter, a visually acceptable square wave can be reproduced from a finite range of frequencies. As the frequency response of a system declines, however, its ability to pass a square-wave image declines proportionately, as does in turn its ability to

Contrast in Modulation-Transfer Functions

Sharpness and Square-Wave Synthesis

FIGURE 3.11
*Square-wave
synthesis from sine
waves*

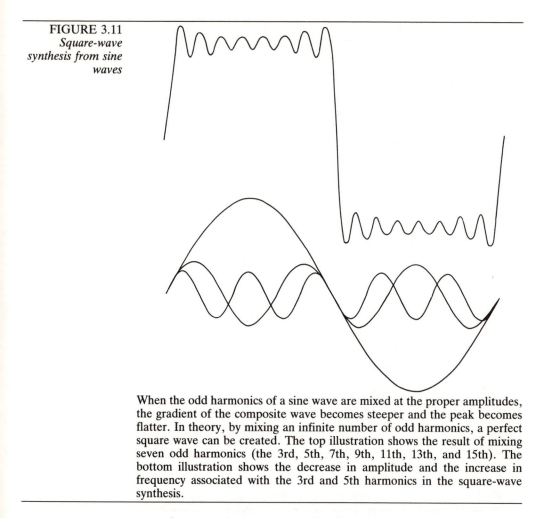

When the odd harmonics of a sine wave are mixed at the proper amplitudes, the gradient of the composite wave becomes steeper and the peak becomes flatter. In theory, by mixing an infinite number of odd harmonics, a perfect square wave can be created. The top illustration shows the result of mixing seven odd harmonics (the 3rd, 5th, 7th, 9th, 11th, 13th, and 15th). The bottom illustration shows the decrease in amplitude and the increase in frequency associated with the 3rd and 5th harmonics in the square-wave synthesis.

reproduce an edge accurately. The sharpness of a system can therefore be surmised from its frequency range as shown on its MTF.

Limitations of the MTF

Although the MTF provides valuable information, a single curve does not completely describe the performance of a lens. A lens' performance will vary according to its specific aperture setting and according to whether it is tested using axial or marginal rays. To understand thoroughly the performance of a lens, one must examine a family of curves generated at different f-stops and different points in the visual field. A drawback, alas, is that the vast amount of information so obtained can easily complicate the issue of comparing one lens with another.

Also, modulation-transfer functions, like acutance ratings, are laboratory measurements. Generating them is all but beyond the reach of the home experimenter. The typical photographer has no convenient way to obtain the modulation-transfer function of a specific component in his or her system.

Information Theory

Information theory is another technique used to evaluate image quality. Like the modulation-transfer function, information theory was first used to evaluate electronic engineering problems. It has been useful in applications ranging from evaluating radar systems to designing electrical transmission lines.

Information theory measures the capacity of a system to store and transmit information. To understand how it applies to photography, think of an image as being composed of several tiny, discrete *elemental areas,* sometimes called picture elements or *pixels,* each having a specific density varying with subject brightness. Photomechanical images like magazine and newspaper pictures are made up in this way. If you examine a newspaper photograph under magnification, you will see that it consists of hundreds of tiny ink dots of various sizes.

Images stored in computers, sent over data-communications lines, or broadcast by television transmitters and satellites are converted to electronic signals before being stored or transmitted. To reduce the cost of handling information and to reduce the demands imposed on communications lines, elements to be stored or transmitted must be as few in number as possible. On the other hand, many pixels are needed to preserve image quality. Several issues arise in the design of an image-transmission system. First, how many elemental areas must be used to achieve a given standard of quality? Next, to what extent can the number of elemental areas be decreased while retaining the identity of the image? Alternatively, how large can individual elemental areas become before image continuity is destroyed? Finally, to what extent can the number of image tones be decreased before the impression of smooth tonality is destroyed? These questions will be addressed here and in subsequent chapters.

To transmit an image electronically, as in television transmissions, the image is scanned line by line in a raster pattern and variations in image density along each scan line are converted to variations in an electrical signal. The size of the scanning beam determines the size of an elemental area. If scan lines are thin and closely spaced, elemental areas will be small and the quality of the transmitted image will be high. As the number of scan lines increases, however, the number of elemental areas to be transmitted increases as well. To handle this added information one must either allow more time to complete the transmission, improve the quality of the communications lines, or increase their number. Both the cost and quality of a transmission system are thus related to its information capacity.

Information is simply an intelligible message made up of *bits,* a bit being the smallest element in a system capable of carrying information. In photography a bit is the smallest segment of an emulsion that can be clear or dense. In high-contrast images, which contain only two levels of image tones, one bit per pixel describes an elemental area. The bit's presence indicates the dense condition and its absence indicates the clear condition.

Bits can be used to represent many different tones if several bits are combined. Two bits together can assume four states (00, 01, 10, or 11) and can represent four different messages or four unique tonal values. Three bits can represent eight different states (000, 001, 010, 011, 100, 101, 110, and 111). The number of distinct messages that can be conveyed by a series of bits increases as a power of 2 in a binary progression with the addition of each bit.

If d represents the number of density levels in an image, the number of bits b needed to represent each density uniquely is given by:

$$b = \log_2 d$$

where the expression $\log_2 d$ represents the binary logarithm of d. The amount of information in an image containing n elemental areas, any of which can assume one of d unique densities, and the capacity required in the transmission channel to transmit such an image, is determined by the relationship:

$$c = n \ (\log_2 d)$$

where c is *information capacity*.

An increase in the information capacity of a photographic system has the potential to increase picture quality. A system with a large information capacity can transmit and store more information about the details and tonalities of an image than can a system with a smaller information capacity.

Information theory provides an alternative way of examining image clarity, defining it in terms of the number of features and tones rendered in an image. This differs somewhat, but not greatly, from the concept of modulation transfer, which considers the size rather than the quantity of image features and their contrast rather than their gradation. These relationships will be clarified in succeeding chapters as you examine in more detail how image clarity is affected by contrast, gradation, and the microstructure of the image.

Chapter 4

The Nature of Image Degradation
Understanding Blurring and Image Softness

By traditional thinking, photographers aspire to improve the quality of their photographs. The practical effect of high-resolution technique is less that of increasing quality as reducing degradation. The difference here is subtle but important. In a strict sense the photographer cannot modify an image to improve its fidelity to the subject. An image can never be more accurate on leaving a reproducing system than it was on entering. If quality is to be added in photography it can be added to the *system,* not to the image, by selecting better cameras, films, and lenses. Having selected the components of a system, one will have established an upper limit beyond which quality can only decline.

It is not enough to know what makes photographs sharp and clear. To improve your skills you must know what makes images blurred. Mastering the craft of photography entails mastering the mechanisms of image degradation.

Several mechanisms are involved here. In fact, every imaging operation, whether to form an image in a lens, record an image on film, or transfer an image from negative to print, is destructive of image information. Most of these mechanisms degrade an image in a similar way, by diverting rays of image light from the path they must follow for precise reproduction of the subject.

Image-Point Spreading

The mechanics of image degradation can be understood by examining how an imaging system renders an isolated image point. If a system is to resolve fine details it must render individual points accurately. Image degradation and the spreading of image light become most evident in images of point size.

In a perfect photograph, if it were possible to create one, each subject point would be reproduced as an infinitesimal image point. Infinitesimal elemental image areas would then yield infinite information content and unlimited resolving power. Needless to say, photographic points are never infinitesimal. The finite size of each point is assured, first and foremost by the wave characteristics of light. Light waves create diffraction patterns that prevent the diameter of an image point from becoming immeasurably small. These diffraction patterns always increase the size of image points and always limit the minuteness of detail that can be rendered by an optical system. Photographic images are further limited by the size of the light-sensitive crystals in film and print materials. These are just the beginning problems of image degradation.

Since a perfectly reproduced image point would have no measurable size or would be infinitely small, any measurable point size gives a numerical indication of the magnitude of degradation in an imaging system. The larger image points are, the more the image has been degraded. The magnitude of image degradation is thus easily measured, for it increases directly as the radius of an image point increases.

Additional information about image degradation can be deduced from the shape of an image point, which offers clues about specific causes of degradation. Certain deformities in image points indicate the presence of lens aberrations, equipment deficiencies, or procedural errors. In fact, an analysis of the shapes of point-sized star images led to an understanding of the mechanics of light diffraction.

In photographs of overexposed line drawings, the damage inflicted by image spreading is visibly evident in the thickening of lines: the heavier the exposure, the thicker the lines. Similar spreading is observed in photographs of fireworks, wherein the thickness of spark tracks is determined by aperture setting. Line thickness in a properly exposed photograph is proportional to line thickness in the original. Such spreading of image features, although exaggerated by overexposure, occurs to some extent in correctly exposed images—it is caused by characteristics of the optical system and of the sensitive emulsion.

Sources of Image-Point Spreading

Sources of image spreading are discussed in detail in Part II. A brief preview of the major sources of point spreading will help you appreciate the scope of the problem.

Optical spreading has various origins. They are grouped for convenience into three categories: diffraction, lens errors, and focusing errors.

Diffraction.   Diffraction occurs in any optical system wherein image-forming light passes through an aperture or close to an opaque edge. The effect that diffraction has on the radius of a point is illustrated in the circular aperture diffraction pattern (see Figure 9.3). Regardless of lens quality, the optical-spread function will never be smaller than the disk created by diffraction.

Lens Errors.   Some of the imaging errors of a lens exist because of its design (see Chapter 14). Design errors or uncorrected, residual aberrations have varied effects on the size and shape of image points. One way to illustrate these effects is by means of spot diagrams. Spot diagrams, which are constructed mathematically from the design equations of the lens, represent the theoretical spot associated with a single image point.

The illustrations in Figure 4.1 show how a point looks when affected by a single aberration. Spot diagrams constructed in this way, however, may not resemble the aberration patterns seen in real lenses. Real image points, being the result of all aberrations acting together, may or may not be as orderly and symmetrical. Coma and astigmatism, in particular, introduce lateral spreading that distorts the shape of image points. Aberrations may also split a single image point into several disconnected, deformed spots.

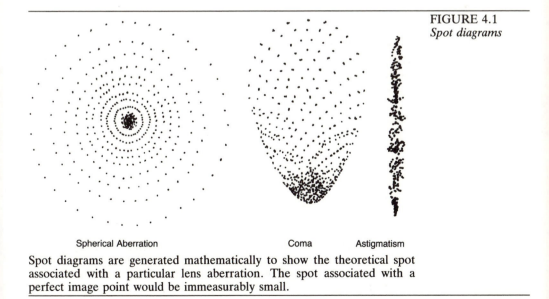

FIGURE 4.1
Spot diagrams

Spherical Aberration   Coma   Astigmatism

Spot diagrams are generated mathematically to show the theoretical spot associated with a particular lens aberration. The spot associated with a perfect image point would be immeasurably small.

Other lens errors occur during construction. It is difficult to mass produce economically lenses of the very highest quality. For example, lens makers cannot consistently mix glass so as to obtain the exact properties specified by the designer: the properties of the mixture invariably differ from batch to batch. Consequently even lenses of the same design perform differently. It is also possible during the assembly of a complex lens to misalign axially one or more lens elements. Such a lens is said to be *tilted* or *decentered*. Unless all lens elements and groups of elements are precisely centered on the optical axis of the lens, the lens will not perform as designed.

Focusing Errors.   Incorrect focusing, inadequate depth of field, misalignment of focusing screens, or uneven pressure exerted by the film pressure plate can all increase the size of an image point and reduce the sharpness of an image. Focusing errors occur because of inaccuracies in adjusting the point of critical focus and because of depth-of-field problems inherent in the geometry of the optical image. The geometry of image formation refers to the process whereby image points are projected as cones of light by a lens. A point can be rendered sharply only when the narrowest part of its image cone falls on the film plane (see Figure 14.7).

Emulsion Spreading

Image spreading occurs in the emulsion of a film, as mentioned in the discussion of acutance in Chapter 3. As seen, the slope of the gradient curve increases in proportion to the spreading of light as it reflects off emulsion crystals to expose crystals outside the path of the original beam.

Spread Functions

Spread functions are devices used to illustrate the spreading and deformity of image points. See Figure 4.2. *Optical-spread functions* describe the distribution of luminous energy across the center of an

FIGURE 4.2
*Density distribution
in an image point*

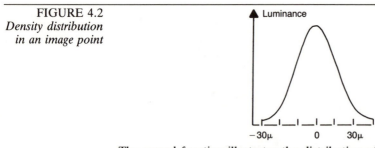

The spread function illustrates the distribution of luminance in an optical point or the distribution of density in an emulsion point. Spread functions are often illustrated using Gaussian curves, the bell-shaped curves associated with the normal-probability distribution function.

optical point. *Emulsion-spread functions* describe the distribution of density in the emulsion of a film across the center of a single image point (the point-spread function) or across a line (the line-spread function). See Figure 4.3. The optical-spread function is affected by the properties of light and by the quality and performance of the lens. The emulsion-spread function is affected by film parameters and by exposure. The *composite-spread function* is a combination of spread functions. It is understood to include all degradation: that which occurs in the negative and in the print, optical spreading from the camera lens and the enlarger lens, chemical spreading during development of films and papers, and blurring from a multitude of sources.

It is the *composite-spread function* that determines how clarity in a photograph will be degraded. The radius of the composite-spread function S, which corresponds to the radius of the smallest image point an imaging system can render, is given by the relationship:

The Composite-Spread Function

$$S = \frac{1}{R}$$

where R is the resolving power of the system. This is an approximation whose origin is discussed at length in Chapter 8. By substituting this approximation into the formula for resolving power

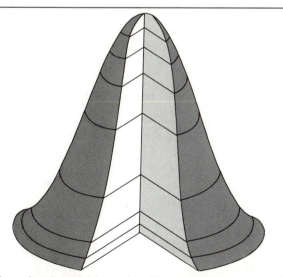

FIGURE 4.3
The point-spread function

A three-dimensional model is used to illustrate the point-spread function. Where the point-spread function is orderly, it can be derived from the line-spread function.

$$\frac{1}{R^2} = \frac{1}{r_o{}^2} + \frac{1}{r_e{}^2}$$

from Chapter 3, it can be seen that the relationship between the composite-spread function and its components is

$$S^2 = s_e{}^2 + s_o{}^2 + \ldots + s_x{}^2$$

where

S = the composite-spread function,
s_e = the emulsion-spread function,
s_o = the optical-spread function, and
s_x = any other component-spread function.

This formula, although it too is an approximation, is thought to show correctly the relationship between the parts and the whole of the composite photographic-spread function. It is a function that increases in size with the addition of each new spread component. Components are independent of one another to the extent that altering the influence of one has no effect on the performance of another. However, the way a component affects the system depends on the state the system is in when that component is introduced. How this happens becomes clearer when actual systems are studied using the equation. This formula is examined in more detail in Chapter 5.

Gaussian Curves

Gaussian curves are often used to illustrate the concept of the spread function and to analyze the effect that point spreading has on image clarity. The Gaussian curve is the curve associated with the normal-probability function.

FIGURE 4.4
Spread functions and image quality

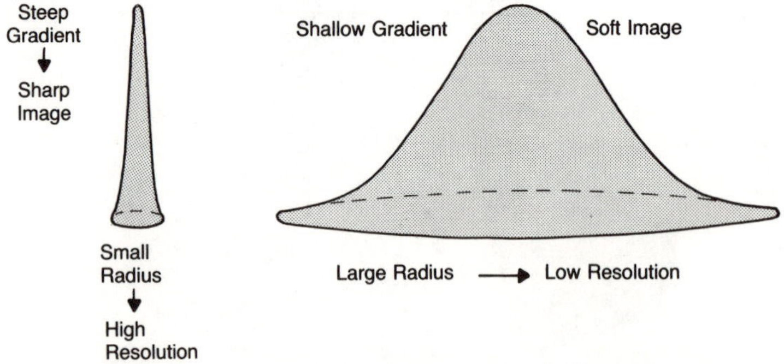

Steep Gradient → Sharp Image

Shallow Gradient   Soft Image

Small Radius → High Resolution

Large Radius ⟶ Low Resolution

Both the radius and the gradient of the spread function, being affected by the same imaging errors, tend to change together. The gradient of the spread function is related to image sharpness; its radius is related to resolution.

Several component-spread functions combine to create the composite-spread function. The distribution of energy or density within some of them is orderly and symmetrical, but within others it is irregular and distorted. Even if the distribution of density within certain component-spread functions is unknown, the density distribution associated with the composite-spread function can be assumed to be Gaussian or normal. This assumption is allowed by the *central limits theorem,* a statistical principle that states that the sum of a large number of independent variables, regardless of their individual distributions, is a random variable whose probability distribution is normal.

This is very convenient. The normal distribution is a fairly simple and well-defined mathematical function that facilitates the study of image degradation. Although the mathematics of the normal distribution will not be explored here, Gaussian curves will be used to illustrate the concept of the spread function and to show how point spreading reduces contrast, acutance, and the resolution of detail. When the spread function of a system is orderly, Gaussian curves allow fairly reliable first-order estimations of how the spread function will degrade an image. It is more difficult to predict from irregular or distorted spread functions how a system will perform. A few examples will demonstrate how useful Gaussian curves can be in analyzing image degradation.

Illustrating Edge Sharpness.   As the composite-spread function expands due to imaging errors and as its gradient flattens, the sharpness of the associated image is reduced. Just as acutance varies with the slope of the gradient curve (see Chapter 3), so does sharpness vary with the slope of the composite-spread function. See Figure 4.4.

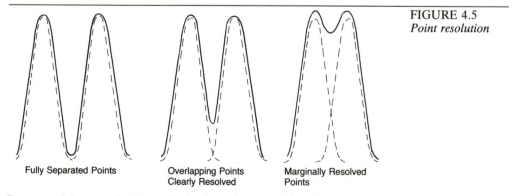

FIGURE 4.5
Point resolution

Fully Separated Points

Overlapping Points
Clearly Resolved

Marginally Resolved
Points

Because of the way density varies in the point-spread function, two points can be resolved even though their spread functions overlap.

Illustrating Point Resolution.   Resolving power depends more on the width or radius of the spread function. Resolving power, you will recall, is based on the separation between adjacent lines or points when they are just resolved. The minimum resolvable separation between image points is affected by two properties of the spread function: its width (see Figure 4.4) and the way its density varies from the center to the edge. See Figure 4.5. One can reliably estimate resolving power from the radius of the spread function alone when its density distribution is simple. Obviously the resolving power of a system will be greatest when the radius of its spread function is small. A small spread function allows image features to be spaced closely without merging and thereby allows very fine image details to be reproduced. The steeper and narrower the spread function becomes, the closer it comes to representing a perfect point. The spread function of a perfectly formed image point would be illustrated by a vertical line without slope and without width. Spread functions associated with high-quality systems thus have small diameters and steep distribution curves.

It is usual to find that when resolving power is high, acutance is also high. You have seen, however, that a system may excel in one property but lag in the other. Spread-function curves offer yet another way to demonstrate how this can come about. See Figure 4.6.

FIGURE 4.6
Sharpness versus resolution

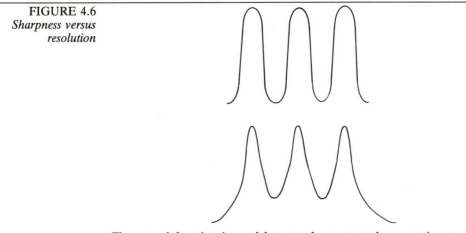

The spread function is used here to demonstrate how one image can be sharper than another yet resolve less detail. In the top curve, image points are large and have steep gradients. A system that produces such points will render an image sharply but with low resolving power. In the bottom curve, image points have sloped gradients but small diameters at their peak density. A system that creates points like these will have greater resolving power than does the previous system, but it will have lower edge sharpness.

The steepness of the spread-function distribution curve also varies with the brightness of a point. The brighter the point, the steeper the curve and the greater its contrast to the background. The significance here is that the greater the contrast or the brightness of a point becomes, the greater its visibility or the more easily it can be resolved.

Note that a loss of contrast causes the shape of the spread function to deviate from its ideal in about the same way as does image spreading. The spread function of a well-defined point has a high height-to-base ratio. Image spreading widens a point causing the height-to-base ratio to decline. Similarly, a loss of contrast reduces the height with respect to the base and the height-to-base ratio again declines. Except for differences in scale, there is little difference in the way that image spreading and contrast degradation alter the spread function. The degradation caused by loss of contrast is very real. Indeed, contrast is essential to visibility; as will be seen repeatedly, a reduction in image contrast can be woefully destructive to the resolution of detail.

Contrast and the Spread Function

Chapter 5

Theory of Image Degradation
Principles of Image Formation

Certain natural laws and principles that explain physical systems seemingly unrelated to photography offer insights into the process of image formation. You have seen, for example, how information theory, originally used to evaluate radios, radar sets, and other communications devices, is also suited for evaluating photographic images. Other principles are examined in this chapter as are some of the logical consequences of the degradation formula introduced in Chapter 4. This chapter will help you understand better why photographs do not always turn out well.

Entropy

It can be demonstrated mathematically that the accuracy of an output signal cannot be improved without reference to the original input signal. It can be shown with equal certainty that the accuracy of a photograph cannot be improved without reference to the original scene. No prospective operation on an image can improve information about a subject once contact with the subject has been severed. Knowledge can only decrease and image quality can proceed in only one direction, that of declining accuracy and clarity. For this reason the imaging process is said to be *irreversible*.

Entropy is a process characterized by a movement or change away from an ordered state toward randomness. A closed system is one that is not acted upon by forces from without. It is a fundamental law of nature that in a closed, irreversible system, entropy must increase. That is, order must decrease and random uniformity ultimately must prevail.

Entropy governs the natural tendency of structured entities to erode, of complex organizations to simplify, of dynamic systems to stabilize, and of information to diminish over time. Entropy is decay. Photographically, an increase in entropy causes contrast to be lost and the resolution of detail to decline. Entropy affects each step in the image-forming process, causing a loss of quality each time an operation is performed on an image. Even as an image is formed in metallic silver, entropy begins to affect its permanence.

An example of entropy in action can be observed in the process of copying a photograph. An original negative and the print made from it are called first- and second-generation images. When a print is photographed, the resulting negative is a third-generation image and the prints made from it are fourth-generation images. Because of entropy, this process of copying and recopying cannot be extended indefinitely; with each successive generation quality is lost and resolving power declines.

Successive Generations

Entropy leads to the conclusion that anytime an image is manipulated or operated on, information will diminish. This conclusion is inescapable. The image, which begins as light waves emanating from the subject, is never better than at its origin; thereafter it is immediately transformed and irreversibly degraded in its passage through apertures and glass and onto films. Any photographic operation that alters the characteristics of an image alters it in the direction of declining resolution. This statement is made in full knowledge of the wonders of electronic image enhancement (see Chapter 7). Image enhancement changes an image so that it is visually more striking or so that available detail is more clearly visible, but it cannot create a more accurate or detailed representation of the subject except by direct reference to the subject itself.

The Inevitability of Degradation

Entropy reinforces what experience teaches: when photographs are made in disregard of high-resolution skills, the deterioration of image quality is a naturally expected outcome. In each step of the photographic process, the degradation of image quality is not only possible but likely. To prevent this loss of quality one must make an active and conscious effort to combat entropy.

The fact that entropy and degradation are inevitable parts of the photographic process reinforces a concept discussed earlier, that control over photographic clarity must lie in the control of image degradation. Image degradation is the operative agent in image formation. It is therefore the agent to be acted upon. The goal of the high-resolution photographer is to combat entropy by reducing degradation, the spreading and deformation of image points, every step of the way.

The Active Agent

The Degradation Equation

The equation

$$S^2 = s_o{}^2 + s_e{}^2 + \ldots + s_x{}^2$$

derived in the previous chapter, is called the degradation equation. This equation predicts the magnitude of the composite-spread function from the separate magnitudes of the component-spread functions. It clarifies the operation of photographic degradation and leads to some important conclusions.

The Accumulation of Degradation

Image degradation is a cumulative function. See Figure 5.1. Resolving power, by way of contrast, is not. All imaging components and image-forming processes introduce an image-degrading spread component. Some of these may be inconsequential in their individual actions, but each, no matter how small, adds to the total and in this way increases the cumulative degradation of the image.

The cumulative nature of the process suggests several ideas about image formation: for example, degradation can accumulate without end and there is no limit to the damage it can do. It also suggests how faulty conclusions can be drawn about matters such as a safe shutter speed, when a single source of blurring is treated in isolation from other kinds of degradation. In analyzing degradation it is not enough that any one source be smaller than the eye can detect. If several component-spread functions are no better than marginally smaller than threshold size, it is likely that the cumulative degradation will be quite visible.

Synergism

The degradation equation can be used to show that image degradation has synergistic properties. *Synergism* means that a system operates as "more than the sum of its parts." The outcome of a synergistic operation can differ from what one would expect, even if the operation of individual components were completely under-

FIGURE 5.1
The degradation formula

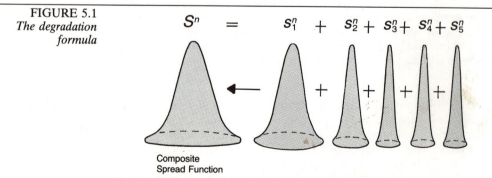

$$S^n = s_1^n + s_2^n + s_3^n + s_4^n + s_5^n$$

Composite
Spread Function

The degradation formula describes the composite-spread function as a cumulative function—it grows with every source of degradation present during the image-forming process, no matter how small.

stood. For example, the change in image clarity caused by introducing a degrading component s_x will depend as much on the magnitude of the composite-spread function S before the component is added, as on the magnitude of s_x itself. In an example cited in Chapter 20, a spread component of fixed size reduces the resolving power of one system by 5 percent and another by 22 percent. This is the basic consequence of synergism. The visible effect a degrading component s_x will have on image clarity cannot be predicted from its value alone.

The Strong-Component Fallacy

The degradation formula also proves that the composite-spread function cannot be as small as its smallest spread component unless all other components are reduced to zero. Some photographers believe that merely by using a super-sharp lens or an ultra-fine-grain film they will get perfectly sharp photographs. This belief is misguided. As a strong component cannot eliminate other independent sources of image degradation, a system will hardly be made perfect by introducing one or two near-perfect components.

Resolution Limits

It can be seen from the degradation equation that the minimum size of the photographic-spread function and the greatest image quality a system can produce will always be limited by the largest component-spread function in the system. When one spread component is conspicuously larger than others, it imposes a resolution limit on the system. The most significant and most immediate improvement in image clarity will be achieved by acting to reduce the limiting spread component. See Figure 5.2.

Say for example that your lens resolves 50 lines per millimeter and your film resolves 25 lines per millimeter. The associated spread functions are 20 microns for the lens and 40 microns for the film. The 40-micron spread function of the film imposes the resolution limit. Should you double the resolving power of your lens or improve some other aspect of your system while ignoring this resolution

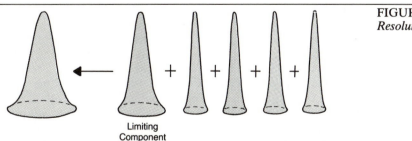

FIGURE 5.2
Resolution limits

Limiting
Component

To obtain significant improvements in image clarity, one must identify the limiting spread component and eliminate it as a first priority. Removing smaller components will have little effect on the image as long as larger components remain active in the system.

limit, you will see little improvement in image quality. The combined spread of the lens and film initially would have been 44 microns; the combined resolving power would have been about 22 lines per millimeter. By doubling the resolving power of the lens to 100 lines per millimeter without improving the resolving power of the film, you will cause the combined spread to drop from 44 to 41 microns; resolving power will rise from 22 to 24 lines per millimeter. Paltry improvements indeed.

Variations in Sensitivity

The degradation equation also illustrates the converse of the previous principle. Introducing a small spread component into a photographic system whose spread function is already large will have little visible effect on visual quality. A system is less sensitive to small changes in image clarity when it contains large spread components. This accounts for the strongly divergent opinions among photographers who have conducted important tests of image quality using their favorite 400-speed film. It does little good to evaluate new photographic components using high-speed, low-resolution emulsions. These films cannot reveal subtle improvements in a product because they themselves introduce too massive a spread component. To evaluate a component or test a procedure reliably, the system used to conduct the test must be of nearly ideal quality or of quality at least as good as that of the component being tested.

Concomitant Reduction of Degradation

The final conclusion to be drawn from the degradation equation is: to maximize the resolution of image detail, all components of the spread function must be minimized simultaneously. Each spread component, without exception, must be eliminated or reduced to its smallest conceivable size. *Everything must be right.* It is a stringent condition, but if it is not met, a photographic system will never yield its peak imaging quality. It is a challenging condition, for there is likely to be some s_x component that precludes your attaining full resolution. Nevertheless, the existence of any spread component, due to any cause, can single-handedly prevent you from getting the best from your system.

From this observation the correct approach to high-resolution photography becomes clear. One must make an unbroken string of smart decisions. In every step of the imaging process, as well as in the steps leading to the actual exposure, decisions affecting the sharpness and detail of a photograph must be the right ones. Certain choices, such as selecting a camera and lens, will already have been made well in advance. Others, such as selecting a film, will be made before each photo session. Aperture and shutter speed must be selected before each exposure, and yet other decisions must be made during processing of the film and print. If in this string of options a single choice is made that compromises image clarity, consummate resolution will be elusive.

Chapter 6

Resolution and Human Vision
How Good Must the Image Be?

Although it may be difficult to capture every nuance of subject detail in a photograph, it may also be unnecessary. The eye cannot resolve every detail in a perfect image just as a camera cannot resolve every detail at the scene. It does not always matter that the recorded image is imperfect. The pertinent question is: how good must an image be to appear sharp and clear to the eye? Conversely, how bad can an image become before it is judged as poor?

Answers to these questions depend in part on the limitations of human vision, on the capacity of the eye to resolve fine detail, and on human tolerance for image degradation and softness. The traditional approach to investigating such matters begins by determining the point-resolution limit of the eye or by measuring the smallest discernible spot; resolving power in lines per millimeter is then computed from the dimensions of this spot. But the connection between the line-pair resolving power of the eye and the size of the smallest visible spot is not what some have assumed it to be. As a matter of fact, this approach has led to difficulties and to errors in thinking. You will see why herein. First, a few terms will be defined.

Point discrimination is one of several ways to describe resolving power. See Figure 6.1. It is a measure of the *minimum visible* size and refers to the smallest spot or the smallest amount of a substance one can observe, that is, of whose presence one can be certain. At the minimum visible size, the image seen is little more than a blur pattern that provides no information about an object other than its location.

A distinction is made between point discrimination and *line discrimination*. The eye responds differently to points and lines. A distinction is also made between line discrimination and *vernier discrimination*—the capacity of the eye to detect a change in line continuity or to determine whether two edges are aligned with or offset from each other.

Two-point discrimination refers to the *minimum separable* distance at which objects are individually resolved—the smallest resolvable angular subtense between two objects. The minimum separable subtense is the angular separation at which one can see that an image represents two discrete objects rather than a single large one.

Shape discrimination begins at the *minimum legible* size. At the minimum legible size one can detect the presence of an object and determine something about its shape and form. *Blending size* is also of interest to this discussion. It refers to the largest pattern size in the substructure of an image that lets it appear that no structural pattern exists.

Point Discrimination

Although not universally agreed upon, there are indications that the smallest object clearly and consistently visible to most people subtends an angle of about one minute of arc (1/60 of one degree or 60 seconds). Smaller objects subtending angles of around one-half minute (1/120th of one degree or 30 seconds) of arc may be discernible at one moment but disappear from view the next.

Whether an object will be resolved at a given moment depends not only on its size but on viewing conditions and on the eyesight of the observer. To say that the point-resolution threshold of the eye is one minute merely means that under good viewing conditions a person with normal vision will see an object clearly and consistently only when its angular size is around one minute or larger. See Figure 6.2.

FIGURE 6.1
Marginal resolution

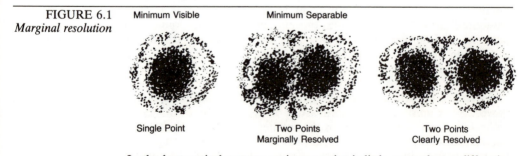

Minimum Visible

Minimum Separable

Single Point

Two Points
Marginally Resolved

Two Points
Clearly Resolved

In the best optical systems an image point is little more than a diffraction pattern. Two points are resolved when the viewer can be certain that the diffraction pattern was caused by two separate points instead of a single large one.

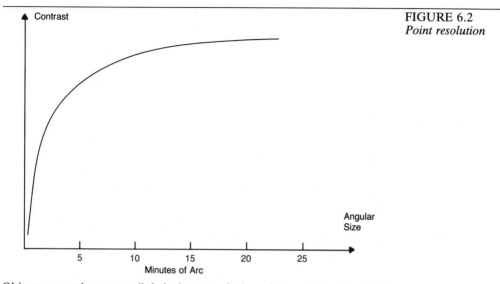

FIGURE 6.2
Point resolution

Objects cannot be seen until their size exceeds the point-resolution threshold of the eye. The eye cannot consistently resolve objects whose angular size is smaller than about one minute of arc.

Of course, the limiting size for point discrimination is not so clearly defined that objects larger than some fixed size will always be seen and objects smaller will not. When it is necessary or convenient to refer to a fixed limit, however, an angular size of one minute, or sixty seconds of arc, will be used.

An angular measurement of point discrimination can be converted to a linear measurement by first establishing the distance at which the linear measurement will be taken. The distance at which vision is best is roughly 25 centimeters or 10 inches. Sometimes one holds an object closer to view it critically, but at closer distances the focusing ability of the eye is impaired. Twenty-five centimeters is a reasonable approximation of the distance used for close visual inspection, reading fine print, and viewing hand-held objects. It will be assumed that the smallest point resolvable by the unaided eye at this distance is the smallest point resolvable.

Viewing Distance

At a given viewing distance D the linear size of an object l is determined from its angular size a using the relationship:

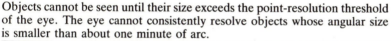

$$l = D\tan a$$

Linear Measurements of Point Discrimination

Since the tangent of one minute of arc is roughly 0.00029, the diameter of the smallest speck visible to the unaided eye at a close viewing distance is about 0.07mm or 70 microns. One can then expect that objects larger than 70 microns will be quite visible when viewed under favorable conditions at the distance at which vision is best.

The Contrast Threshold

Before an object can actually be seen, it must exceed a second visual threshold. It must differ from its background in brightness or in color. Under the very best viewing conditions it must differ in brightness from adjacent areas by about 1 percent. In poor lighting it must differ by 5 percent or more. This visual limitation is known as the *contrast threshold*.

The two thresholds of vision, though different, are mutually dependent under certain conditions. This dependence shows up in one way when the size of the test object is near the threshold size. At angular sizes just greater than 60 seconds, an object's brightness must differ from that of the background by about 20 percent if it is to be clearly visible.

As contrast increases, however, objects of increasingly smaller size become visible. A bright object or a point-sized light viewed against a dark background (a bright star, for example) can be seen at angular sizes smaller than one minute. Apparently, if an object is bright enough, it remains visible at any angular size.

Characteristics of Vision

Measurements of visual acuity are consistent with what is known about the structure of the eye. The light-sensitive membrane of the eye, the retina, is a collection of neuroepithelial cells with one of two kinds of terminal light receptors: rods, which are responsible for night vision, and cones, which are the principal receptors for normal daylight vision. A relatively small part of the retina, the fovea, is located directly in line with the axis of the lens at the center of the field of vision and has a very high concentration of cones, about half a million per square millimeter. It is responsible for high-resolution vision. The separation between cones in the fovea is on the order of 0.0015mm, which equates to an angular separation between cone centers of approximately 20 seconds of arc. This value is close to the one just discussed as the point-resolution threshold of vision. If one considers that the resolving power of the eye cannot be quite as good as 20 seconds because of aberrations in its lens, then 30-60 seconds again seems to be a fair estimate of the point–resolution threshold of human vision.

Line Discrimination

There is a substantial difference in the size of the minimum visible line and the size of the minimum visible point. For points to become visible, for the brain to distinguish between random cone excitations and actual visual stimuli, an object must project a large enough image on the retina to excite several cones simultaneously. Since cone centers are separated by an angular distance of 20 seconds, the diameter of a point image must exceed 20 seconds by a substantial margin for that image always to stimulate two or more cone centers.

Lines too must stimulate several cones to become visible, but lines can stimulate cones all along the length of the retinal image. If a line is long enough it can stimulate several cone centers even though the line itself is narrower than 20 seconds.

Indeed, empirical measurements of visual acuity show that a dark line viewed against a bright background can be resolved at a width of one-half to one second of arc. From a close viewing distance this is a linear size of about one micron or less. Compare this to 70 microns, the size of the minimum visible point. The visibility of a bright line on a dark background is limited only by its brightness, as is the visibility of a bright point. Lines smaller than one micron can be seen if they are made bright enough.

The improved performance of the eye in resolving line images seems to apply, however, only to lines isolated in space. The visual process that resolves closely spaced details apparently differs from the one that resolves single lines; the outcome is far less spectacular. Closely spaced bright lines on a dark background must be separated by about one minute to be seen as separate lines. Dark lines on a bright background can be resolved at separations of about 40 seconds.

It is easy to misunderstand the relationship between the line-pair resolving power of the eye and the point resolving power of the eye. A common error has been in misinterpreting the resolving-power formula:

The Resolving Power of the Eye

$$R = \frac{1}{d}$$

In this relationship, *d* represents the *separation* between lines from center to center at the resolution limit; it has sometimes been thought to represent the diameter of a resolved point or the thickness of a target line. It happens that standard resolving-power test targets are designed so that lines and spaces are of equal width. When they are, the distance from the centers of two adjacent lines equals the width of a line plus a space or twice the width of a line. The correct value for *d* is likewise twice the width of a line.

Also, using the *diameter* of the spread function as the value of *d* is a mistake. Points made up of ink dots, which cannot be resolved individually when they overlap, must be physically separated to be separately detected. Points created by spread functions, because of their gradual change in density, can overlap and be resolved. Consequently the diameter of the spread function is not equal to the width of a marginally resolved line in a resolving-power test target, nor is it equal to the width of a line-space pair. This observation is easily confirmed. Starting with a system of known resolving power measured in line pairs, direct measurements of that system's circle of confusion or spread function reveal that the diameter of an isolated point is greater than the thickness of a marginally resolved test-target line by a factor of about four (see Figure 8.2).

A misapplication of the resolving-power formula has caused differences in traditional estimates of the best quality the eye can appreciate at a close viewing distance, differences ranging from 5 to 25 lines per millimeter. In some such cases, the line-pair resolving power of the eye has been computed using the minimum visible instead of the minimum separable; that is, using object size instead of object separation. Taking 70 microns as the width of a resolved line, 5 to 7 lines per millimeter may be thought to be about as much detail as the eye can appreciate. But when the value for d is determined as it should be, by the separation between marginally resolved image features, which is shown by measurements of visual acuity to range from 40 seconds to one minute, the line-pair resolving power of the eye would seem to range from 14 to 20 line pairs per millimeter. It appears in fact that there are perceivable differences in quality among images resolving up to about 25 lines per millimeter. See Figure 6.3.

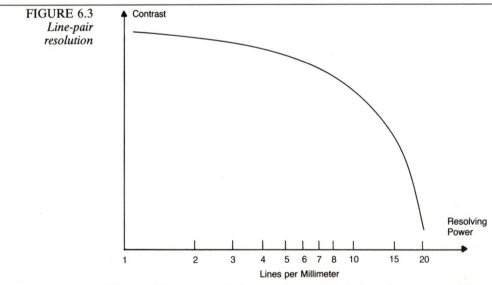

FIGURE 6.3
*Line-pair
resolution*

The resolving power of the eye does not end abruptly at some arbitrary point—it tapers off. Lines are just marginally resolved at line spacings between 10 and 20 lines per millimeter.

Blending size is a function of both image structure and viewing distance. Although the microstructure of a photograph becomes visible when the image is greatly enlarged, the evidence of degradation lessens with distance so that, from far enough away, blurred image features seem sharp and clear, and broken tones seem smooth and continuous. Increasing the viewing distance so improves the visual appearance of an image that, when viewed from far enough away, even images of poor quality seem quite clear. Greatly enlarged motion pictures, when viewed at arm's length, would seem coarse, grainy, and blurred. The quality of a motion-picture image is acceptable because the picture is seen from the middle or rear of the theater. Huge billboard images are successful as well because of the great distance from which they are viewed.

The size of the spread function when signs of degradation and graininess disappear in a photograph is the *blending size* for that photograph at that distance. Only when the structural details of an image are smaller than the blending size of vision is the image as good as can be appreciated.

Blending Size and Blending Distance

Minimum Acceptable Resolving Power

There is a difference, a great one, between the resolving power of the best image appreciated and the worst image tolerated. An image that resolves 25 lines per millimeter contains more detail than the unassisted eye can ever see; improvements to such an image go unappreciated. Below 25 lines per millimeter one begins to sense that an image has structure. At slightly lower levels of resolution one can see that structure, yet sharpness and detail are acceptable. How much can resolving power decline before an image is perceived as poor in quality? To find out, let's examine the structure of photomechanical images.

You may agree that the photo reproductions in a first-rate magazine are very good. Yet when you look closely at them and know what to look for, you can see the ink dots used to create these images; they are larger than the resolution limit of the eye. There is something to be learned from the way the eye responds to these dots.

Bear in mind that the resolving power of a photomechanical image depends more on the size of an elemental area than on the size of individual dots. The size of an elemental area is determined by the size of the screen used to make the halftone plate. Ink dots vary in size depending on image density at their location, but elemental areas have the same size throughout the image.

When screen size is given in dots or lines per inch, the limiting resolving power in lines per millimeter can be determined by directly dividing dots per inch by 25.4. A 150-line-per-inch screen, for example, has a limiting resolving power of about 6 lines per millimeter. Resolving power can be worse with such a screen, but it can never be better.

Newspaper photographs made with a coarse, 50-line-per-inch screen (2 lines per millimeter) are not as good as magazine photographs, yet newsprint pictures are often thought to be quite clear and sharp. Such images prove that the structural features of an image do not have to be smaller than the resolution threshold of vision for the image to be judged of good quality. Similarly, the photographic-spread function does not have to be smaller than the visual-threshold size for a photograph to appear sharp and clear. Human tolerance for low-resolution newsprint images suggests that the resolution of fine detail is not the only factor that determines whether an image will be judged good or bad.

Toleration for Image Degradation

The usefulness of a visual stimulus depends much on whether it helps one to recognize subject matter. The resolution of fine detail, though it gives valuable clues, is not always essential to this recognition. One can identify a human form even if a photograph does not resolve skin pores, and one can identify a garment even if a photograph does not resolve its fibers. The absence of fine detail rarely renders an image fully useless.

An image becomes meaningless when dominant features of the subject cannot be made out or the subject cannot be readily identified. Halftone images break down, for example, when dots are so large that image patterns are not apparent—the dots cannot be mentally assimilated into a useful image. That is, they cannot be organized by the brain into a coherent pattern containing meaningful visual information. They appear as random spots with no obvious connection. It is not until dominant image patterns reach this state of incoherence that the image becomes unacceptable as a source of visual information. An image must virtually disintegrate before human tolerance for image degradation is exceeded.

Elementary Image Recognition

An important biological demand placed on vision is that it allow danger to be perceived quickly. In this regard, visual recognition satisfies instinctive protective mechanisms. There is no corresponding need for detail and resolution. To the contrary, one learns to expect that visual detail will be lost as one moves farther from an object or as the object gets smaller. The loss of clarity here is rarely objectionable, even at the original scene. That which is not seen clearly is assumed to be either too remote or too small to pose an imminent threat. Likewise, photographic images of low resolving power can be tolerated if large, ominous objects, the major forms in the image, can be recognized.

Tolerance for loss of sharpness, as opposed to loss of detail, is another matter. In real life sharpness does not diminish with distance or size. Edges remain visually sharp from near to far, so one expects to see sharp edges in a photograph. Blurred, fuzzy edges are bothersome. Loss of sharpness more than loss of detail is the reason low-resolution images are objectionable.

Chapter 7

Image Microstructure
A Close Look at the Photographic Image

Specific image properties are known to create the visual impression of detail and sharpness. The presence of fine detail—tiny shadows and highlights—confirms that the image is one of good resolution. The presence of high edge contrast confirms that the image is sharp. It is known also that image clarity can be degraded by either of two mechanisms: by point spreading or by a reduction in point contrast.

How these mechanisms operate within the microstructure of the photographic image is the subject of this chapter. Examined herein is the silver image, the likeness formed in the emulsion layer of the photographic material, and the optical image, the likeness formed by light. The chapter ends with a discussion of ways that the microstructure of an image can be modified to enhance image appearance.

The Silver Image

Although new details may be visible in a photograph as it is enlarged, the improvement does not continue indefinitely. Even after slight magnification, say of 10 to 20 diameters, image detail declines. Part of the reason is that photographs are *discrete* images as opposed to the *continuous* images of nature. They have a finite structure composed of observable elements.

The image in a photograph consists of particles of pure elemental silver which were once thought to be solid spherical masses. The powerful electron microscope revealed them to be intertwining strands resembling loosely packed fibrous balls. Silver is apparently extruded from development sites in filaments. Large crystals produce many such filaments, but the very smallest crystals produce just a single strand. The dense areas of a photograph or negative consist of millions of extremely fine silver filaments of just a few atoms in diameter. When they intertwine and overlap to a depth of several microns, they are able to form a dense opaque blanket.

Graininess

Photographs of the very highest quality seem to have a continuous structure, but upon magnification it becomes evident that even good photographs have a granular structure. Even at a comparatively low magnification ratio, any photograph takes on a mealy, mottled, grainy appearance.

It is tempting to suppose that the grainy particles seen in an image at moderate magnification are the individual silver particles associated with a single emulsion crystal, but they are not. It takes a powerful microscope to see the silver specks directly. Graininess is caused by a complex visual mechanism arising from the random and irregular spacing of the individual silver particles and the apparent clustering of the original silver-halide crystals. Randomness is an important part of the visual effect of graininess. When image elements are spaced in a regular pattern the visual sensation of graininess is lessened.

It is difficult to generalize mathematically about the extent that grain reduces resolving power. One can be certain though that when image features are as small as the grain pattern, they cannot be distinguished from the grain itself and will technically be unresolvable. Graininess, which is most serious in fast, high-speed films, places a major limitation on the imaging quality of such films. It reduces their resolving power to such an extent that most other sources of image spreading are subordinated. Refinements in high-resolution technique are of little effect in overcoming the degradation of resolution caused by the graininess of the very popular high-speed films.

Grain and Image Points

It may also be tempting to suppose that the grainy specks in a photograph correspond to image points. Of course, they do not. The grain of a fine-grain film may turn out to be just a fraction of the size of the blur spot associated with an image point. The composite-spread function is the parameter that determines the size of the smallest image point rendered in a photograph; grain and other aspects of the emulsion-spread function are merely components of this composite-spread function. See Figure 7.1.

The Optical Image

To grasp fully how an image is reproduced and degraded in a photograph, one must also examine the structure of the optical image. Keep in mind that this optical image is overlaid upon the granular structure of the emulsion so that the final image is degraded by all image defects, whether these defects lie in the silver structure or in the optical pattern. First, we will consider how an optical image is formed, then we will see what is meant by structure with regard to an optical image.

An optical image starts out as rays of light reflected from or emitted by the subject. When these rays pass through a small opening or aperture, an image is formed in space called an *aerial image.*

A lens is not essential to the image-forming process—a fact proven by the existence of pinhole cameras. Lenses are added to imaging systems as a refinement. Without them, small apertures are needed to reduce the size of the blur spot and to obtain reasonable image quality. Yet when apertures are small, not only does the image grow dim, but the diffraction disk grows larger and neutralizes much of the expected improvement.

A lens, by converging light rays, allows use of large apertures. The convergence of the lens reduces the blur spot; the large aperture reduces the degradation caused by diffraction and creates a brighter image. Reducing the size of the focal spot is the essential function of a lens. Some lenses do this better than others, but no practical photographic lens can bring all the rays of a single image point into convergence precisely at the same location.

When light rays enter a lens they are diffracted and diverted by lens errors from their ideal paths to create a patch of light instead of a well-defined point. This patch is called the optical-blur circle and is essentially the focal spot corresponding to the optical-spread function. An optical image is composed of many blur circles, one associated with each point of the subject. But at the image plane, the enlarged blur circle associated with an image point overlaps blur circles of several adjacent image points. The resulting optical image

The Optical- Blur Circle

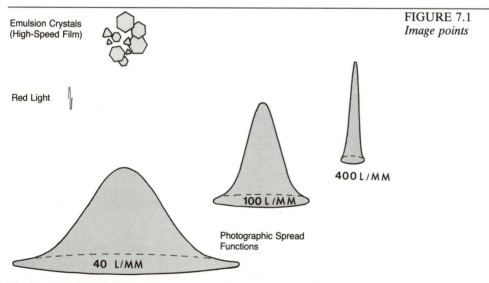

FIGURE 7.1
Image points

The smallest detail resolved in a photograph may be limited by grain or by other aspects of the composite-spread function. In any case it is the composite-spread function, not grain, that corresponds to an image point.

thereby consists of *overlapping blur circles* wherein image brightness at any given image position is determined by the average brightness of all blur circles overlapping at that position.

Point Spreading and the Resolution of Detail

One serious consequence of image-point spreading is that it increases the extent to which adjacent image points overlap, and it is this overlapping of points that lowers the visibility of fine image features. When contrasting adjacent points spread and merge with one another, the contrast between them is weakened. In the presence of image spreading, small subject features, those rendered by few contiguous image points, are overlapped by so many background points that they take on the characteristics of the background and become indistinguishable from it. They vanish. The overlapping of image points thus weakens elemental contrast throughout the image.

It is by this mechanism, by the degradation of contrast within the microstructure of the image, that image spreading reduces the resolution of detail. And it is through this mechanism that contrast, the spread function, and the resolution of detail are related. In a system that has substantial point spreading, a subject feature must differ greatly in brightness from surrounding points if it is to remain visible in the image, or it must be large enough to create many contiguous overlapping image points so that image density is not completely diluted by overlapping background points. The larger the spread function of the system, the larger a subject feature must be before its image density rises substantially above that of the background. As image spreading worsens, only the largest and most contrasting features of an object remain visible.

Blur Circles and Pointlights

The optical image thus has structure, albeit a subtle one. Because of the way blur circles overlap, this structure is not ordinarily seen; optical images seem continuous even when the spread function is large. One will not see individual image points in an optical image except when the points are of very high contrast. Small specular highlights, catchlights, or pointlights, as they are variously called, are sometimes of point size. If so, one can determine from them something about the optical structure of the image, about degradation, or about the size and shape of the spread function. If pointlights are small it is reasonably certain that the system is of high quality and that image detail has been accurately recorded.

Characteristic Blur Patterns.   If pointlights are large, the image has been degraded. If they are deformed, one can often guess from their shape the cause of the degradation. Blurring from image motion has a characteristic streaking pattern which renders pointlights as lines instead of disks; the length of the lines is proportional to the amount of motion at the film plane. Camera motion creates linear streaking of pointlights in the direction of motion. Rotational camera motion creates circular streaks that worsen away from the center of rotation.

Patterns of double lines indicate that the camera may have been jarred, perhaps by the shutter or by the motion of the viewing mirror. Enlarged pointlights that take the shape of the diaphragm—octagonal in the case of certain leaf diaphragms, doughnut shaped in the case of catadioptic telephoto lenses—indicate that the image may not have been in critical focus in that subject plane. Pointlights outlined in a rainbow of spectral color are affected by one of the chromatic aberrations.

Image Enhancement

The microstructure of an image can be altered with the aid of computers in ways that compensate for point spreading, graininess, and degraded contrast. In this process, known as image enhancement, sensitive optical instruments scan the image in small segments to measure the density and chromatic characteristics of individual elemental areas. The minimum size of an elemental area and thus the resolution of the scanning process is determined by the size of the scanning aperture. The density and chromatic characteristics of each elemental area, after being converted by an analog-to-digital process into numerical values, are fed into a computer, which manipulates the data and reconstructs an enhanced image according to programmed instructions.

If the purpose of an enhancement is to reduce grain, the computer essentially reproduces each segment of the image to make it internally homogeneous. When a segment contains the gaps and spaces associated with grain, the computer substitutes a smooth and continuous tonality either by assigning to the segment the characteristics of a dominant adjacent tone or by taking the average of tones within the area itself and assigning that value to the whole.

Enhancements made in this way are likely to obliterate some of the fine detail in an image because the size of the area manipulated by the computer must be larger than the grain pattern of the film; otherwise the computer will be unable to distinguish between grain and legitimate image gaps and spaces. This increase in the size of the elemental area is likely to reduce resolving power somewhat. Small image features are consumed during the averaging process.

Sharpness also can be improved by computer enhancement. In this case, the computer is given a statistical approximation of the spreading within image points and is programmed to modify elemental areas to compensate for the original spread component. Edges can be reconstructed with excellent sharpness using this technique.

It is important to realize that images enhanced in this way do not contain new information. They contain no image details other than those recorded in the image before the enhancement. The sharpness brought to the image may seem like new information, but this sharpness is based on statistical approximations of missing information, not on new knowledge of the subject.

Other approaches to image enhancement use contrast expansion, color transformation, or both to improve visible contrast in subject features. These methods have been known to operate at the expense of resolution and brightness, but when combined with sophisticated analog or digital-processing techniques, they are capable of bringing out the greatest possible clarity of surface features.

Enhancements are properly compared to an artist's reconstruction. They may or may not create a more accurate representation of the original scene. Nevertheless, computer enhancement is of great value in many areas of investigation. Enhancements often clarify the relationship between image components and speed up the visual assimilation of information in the image. Just be on guard—one cannot be certain that everything seen in an enhanced photograph actually appeared at the original scene.

Chapter 8

Measuring Image Degradation and Quality
Communicating Numerically about Blurring and Clarity

An important concept in high-resolution photography is one that by now should seem obvious: high-quality images are achieved by reducing degradation. The larger part of skill in high-resolution photography involves detecting and manipulating troublesome sources of image spreading. In applying this concept one must consider that large sources of degradation must be eliminated first; their presence makes it difficult or impossible for one to isolate and correct lesser sources. Practicing photographers who develop an instinctive feel about the size of various spread components will therefore have an advantage over those who do not. They will be able to sense which sources of degradation are active and will be able to negate them in order of their magnitudes.

This chapter introduces several simple methods whereby spread components may be compared and ranked. It shows not only how to measure image quality, but how to measure the magnitude of image degradation and how to predict the performance of photographic systems. It shows first how to measure line-pair resolving power. It then shows how resolving-power ratings can be used to derive other methods for evaluating image clarity and degradation.

Using resolving power as a point of reference in this chapter is an expedient to using more objective measurements like acutance or modulation-transfer functions. Gradient curves and modulation-transfer functions are of great importance in laboratory studies, but specialized equipment is needed to measure them. They offer accurate and reliable indications of image clarity, but absolute rigor is not always needed by working photographers in routine tests, especially when the test involves a direct comparison of one system with another. In such comparisons the intent is merely to determine which system is better. When one person rates both systems, the objectivity of gradient curves and the modulation-transfer function becomes irrelevant. Resolving-power ratings work well here for the use to be made of them, and they are easily measured by the typical photographer.

There is actually little need to defend resolving-power ratings. One of the toughest tests of a communication channel or of a photographic system is in its ability to pass a square wave—a square wave represents an infinite frequency band that can only be passed accurately and completely in a perfect system (see Chapter 3). A photographic system that can accurately reproduce a square-wave input, such as the bar targets of a resolving-power test chart, can reproduce any image, including one having the finest detail, without difficulty. Resolving-power tests are fairly rigorous.

Measuring Resolving Power

As you already know from Chapter 3, resolving power is determined from the number of lines rendered marginally but clearly visible by a system or a component. It is measured using a special test chart containing bar-space targets arrayed in various patterns and sizes. The first step in the measurement is to photograph the chart or, if the resolving power of a lens alone is being measured, to view the chart directly through the lens. From the photograph of the test chart or from the optical image, find the smallest target group and pattern in which lines and spaces are separately and individually resolved; if necessary, magnify the image to find it. The resolved target need not be imaged with perfect sharpness and clarity; it is only necessary that each bar in the target be separately visible and that a distinct change in density occur between each bar. If no pattern is resolved the test should be repeated at a closer distance. If all targets are resolved the test chart should be moved farther from the camera and the test repeated. Once a marginally resolved pattern is found, use the instructions given with the test chart or use the formula below to determine resolving power.

$$R = \frac{u}{Fd}$$

where

R = resolving power,
u = the distance from the lens to the test chart,

F = the focal length of the camera lens, and
d = the line-pair width of the smallest resolved pattern.

Resolving power can be measured either under *laboratory* conditions, where most factors that influence the outcome of the test are controlled by the photographer, or under *field* conditions, where the hazards encountered on a typical assignment can change the outcome.

Laboratory conditions are appropriate for testing a lens, film, or other component in isolation from the rest of the system where one component is to be compared with another. Remember that tests conducted under controlled conditions cannot predict how a system will perform on assignment. Laboratory tests often yield deceptively better resolving power than can be achieved in fast-paced field situations.

As a practical matter, photographers are more often concerned about picture quality. They want to know how degradation will alter the *image*. They want to know what to expect in actual photographs made using their particular techniques and their system. *System* means the entire package of equipment and skills used to make a photograph. System performance is best determined by *field testing* so that the test includes the effect of the lens, the film, the film developer, the enlarger, microvibrations, and anything else that can lower image quality.

For these reasons you may want to measure resolving power using the equipment and procedures you will use in the field. If you intend to use your system most often outdoors without a tripod, perform the test outdoors without a tripod. If you intend to use supplementary lights or flash units regularly, use them in your resolving-power tests. If you plan to operate in several modes, conduct a separate test in each important mode. The idea is not to determine the peak resolving power of the system, but to determine how much resolving power you can expect under typical working conditions.

Image Legibility

Resolving-power ratings computed from field tests can be used to predict the picture quality or *legibility* of enlarged prints. Legibility usually refers to the readability of written characters, documents, or printed matter. But the clarity of textural material and of continuous tone photographs are closely related concepts; both depend on the resolving power and sharpness of the imaging system. Legibility, used here to indicate the visual quality of photographs, is defined as the number of lines per millimeter resolved in an image at its final enlargement size. As such, photographic legibility is an indication of the quality of an enlarged print.

From what has been said in Chapter 6 about the best image the eye can appreciate, it is clear that image legibility of 25 lines per millimeter represents a very high standard of clarity. Visual quality remains good, however, even after legibility declines substantially

TABLE 8.1 *Image legibility*	Resolving power	Legibility	Image quality
	8	Excellent	Degradation is barely detectable; images are of high quality; clarity is excellent
	4	Good	Degradation is visible but does not affect image recognition; written material is clear enough for prolonged study
	2	Decipherable	Degradation is pronounced; the identity of letters is doubtful; words can be read, but many must be guessed from context; large shapes can be identified
	1	Poor	Images are of acceptable quality only when viewed from a distance; writing is illegible

At a close viewing distance, about 25cm.

from this level. Indeed, image legibility as low as 6 lines per millimeter is judged by some viewers as excellent and of exhibition quality.

The acceptable standard of legibility varies, though, so you may wish to maintain a higher or lower legibility in your prints. Table 8.1 will help you decide. In establishing a standard, consider that the reprographics industry finds 4 lines per millimeter to be adequately legible in noncritical microfilm reproductions. Consider also that the apparent or visible quality of an image is affected both by its legible detail and by viewing distance. If the image will be viewed at a great distance you can adopt a lower standard of legibility without reducing visual clarity.

Enlargement Latitude

As the enlargement ratio of the image or as print size increases, image legibility declines. The enlargement latitude E of a negative or transparency depends on the resolution of detail R in the original and on the minimum acceptable legibility L in an enlarged reproduction:

$$E = \frac{R}{L}$$

This relationship shows how enlargement latitude declines with reductions in resolving power. If prints must retain a legibility rating of 4, a negative resolving 40 lines per millimeter cannot be enlarged by more than 10 diameters without exceeding this limit.

Enlargement latitude increases, however, as expectations of visual quality in the enlargement decrease. For example, one can enlarge the same negative up to 20 times by accepting a legibility of 2. When 8 × 10-inch prints are made from 35mm negatives (an enlargement ratio of about 8 times) a system resolving 50 lines per millimeter will provide a legibility of 6 if the negative is not cropped in printing. A system that resolves 30 lines per millimeter or slightly more, will give 8 × 10-inch enlargements with legible detail of about 4 or less.

Although there is no theoretical limit on how much one can enlarge a 35mm negative, there is a limit to how much quality can be retained in oversized enlargements. Legibility is traded for image size. If mural-sized prints are needed and this tradeoff cannot be allowed, one has no alternative but to use a high-resolution system.

Reduction Latitude

In a similar way one can determine the extent to which printed material or artwork can be reduced in size in a negative so that the quality of the original is retained when the image is enlarged back to its original size. An ideal microcopy can be enlarged to size without a detectable loss in sharpness, but copies need not have perfect definition to be usable. The greatest reduction or minification M obtainable from a system having resolving power R such that the enlarged image retains a legibility rating L is given by:

$$M = \frac{R}{L}$$

The microfilm industry gets extraordinary quality out of the photographic process, routinely achieving reduction ratios of 48 to 96 diameters. The reductions achieved through integrated circuit technology have been even more astounding, exceeding 200 diameters with legible copy. These statistics are mentioned to show the ultimate capabilities of photographic materials and components. Quality of this order is about the best attainable using silver-halide processes. Such levels of reduction can be achieved only using diffraction-limited optics and films of extremely high resolving power. By contrast, ordinary cameras and ISO 1000-speed films usually cannot achieve reductions greater than 10 to 15 diameters at usual standards of legibility.

Spread Radius

Noted earlier was the importance of gaining an instinctive feel about the magnitude of various spread components. One way to do so quickly is to compute *spread radius*, an estimate of the radius of a single image point, or the radius of the spread function, derived from resolving-power ratings.

Such an estimate presupposes a mathematical relationship between line-pair resolving power and the size of the smallest visible image point. The relationship between the resolution of isolated lines and the resolution of isolated points, as noted in Chapter 6, is not what one might expect. It seems though that when closely spaced points and closely spaced lines are marginally resolved, the separation between them is about the same. Assuming this to be so, the discussion will proceed on the notion that as the radius of the spread function increases, line-pair resolving power declines. When the resolving power of a system is given by R, the spread function S must be inversely related to R. The relationship sought will take the form $1/R = d$, where d, the separation between resolved features, is replaced by an expression containing S, the spread function. One possibility is:

$$\frac{1}{R} = kS$$

where $kS = d$. In this relationship k is apparently related in some manner to the separation between the centers of two image points when the points are marginally resolved.

In a diffraction-limited system the separation between resolved points can be established by applying Rayleigh's criterion for point resolution. See Figure 8.1. This criterion asserts that two points are resolvable when their diffraction patterns are separated by a distance equal to the radius of the central diffraction disk, also called the Airy disk (see Chapter 9). In a diffraction-limited optical system, the Airy disk and the spread function are one and the same, so the separation between marginally resolved image points d equals the radius of the spread function S. Since $kS = d$, the scaling factor k reduces to one when S is measured in units equal to its radius. The relationship between the spread function and resolving power in a diffraction-limited system simplifies to:

$$\frac{1}{R} = S.$$

FIGURE 8.1
Rayleigh's criterion for two-point resolution

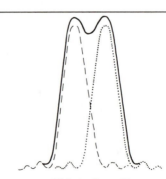

Minimum Separable

Lord Rayleigh asserted that in a diffraction-limited system the diffraction patterns of two points will be resolved when the first minimum of one falls on the central maximum of the other. In other words, two points are resolved when separated by the radius of the Airy disk.

Although this relationship applies to a diffraction-limited system, there are reasons to believe that it will work with spread functions in general. If so, the radius of the composite spread function S of a system can be estimated from the system's resolving power R using

$$S = \frac{1}{R}.$$

It is important to remember that this relationship is approximate. Its accuracy depends on the validity of Rayleigh's criterion as well as on the accuracy of the liberal assumptions made herein about the nature of photographic-spread functions. It is also important to remember that the *radius* of the spread function must be used in this equation—not its diameter.

Spread radius, at last, provides a means for evaluating image degradation in numerical terms and putting it into perspective. See Figure 8.2. What, for example, is the effect of adding one micron of spread to the radius of image points? A micron, being a millionth of a meter, seems too small a quantity to be considered. Yet the micron turns out to be of a convenient size for measuring degradation. Image spread of mere microns can cause considerable damage to image quality.

Suppose your system has the potential to resolve 50 lines per millimeter and your goal is to get all the quality the system can deliver. To do so, you must limit the radius of the spread function to just 20 microns. This is as large as the composite-spread function,

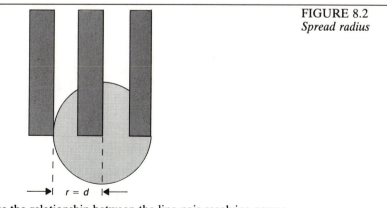

FIGURE 8.2
Spread radius

This figure illustrates the relationship between the line-pair resolving power of a system and its spread function. The separation d between marginally resolved lines corresponds approximately with the radius r of the spread function. This relationship allows estimates of the magnitude of image degradation, represented by the radius of point spreading, to be computed from line-pair resolving-power measurements.

the accumulation of all image spreading, can become. Each component-spread function must be smaller than 20 microns—smaller by a significant margin. Even with an excellent lens and fine-grain film, the optical-spread function, because of diffraction and aberrations, will make up a large part of the 20 microns, and the emulsion-spread function will make up a large part of the remainder. If the film and lens each resolve a respectable 100 lines per millimeter, for example, they will introduce 10 microns of spread each. Using the degradation equation in Chapter 5, it can be seen that their combined total will be somewhere around 14 microns. There will be little room for any additional image spreading, even of only a few microns.

In view of the magnitude of various mechanical sources of image spreading, displacement-spread components arising from the mechanical sources have a potential to do enormous damage to the image if they are allowed to act on the camera while the image is being recorded on film. In ordinary activities a motion as small as 0.2mm, such as that caused by the beating of the heart, can escape detection, but such motion can destroy an image if it registers on film. It equates to 200 microns of linear spread—10 times the limit in the example just cited. Mechanical image displacement this large can nullify the good done by other high-resolution techniques. Image motion, as will be emphasized again and again, even in small amounts, makes high-resolution results difficult to achieve.

Part II

Mechanisms of Clarity and Degradation

Ideally a photograph should be a point-for-point projection of its subject where, for each point on the subject, there is an image point whose size, shape, intensity, and color matches that of the parent point. This can never be made to happen in a photo-optical system. Why? Because whenever an image passes through an aperture of finite size, image points are degraded by diffraction. Because whenever light passes through glass from a medium of a different refractive index, it disperses into a spectrum. Because lenses with spherical surfaces cannot bring image rays precisely to a common focus. Because photographic films reflect and refract light within the emulsion. Because during development, image silver extends beyond the boundaries of original emulsion crystals. Because photographs of three-dimensional subjects simply cannot be made in which the entire subject is rendered with critical sharpness. And because obtaining absolute immobility in a camera during exposure can be difficult or impossible.

An examination of the ways in which these image-degrading factors operate reveals that, while their effects can often be minimized, they cannot be totally eliminated. The challenge is not so much to prevent degradation, as to reduce it and to know how much of it must be accepted due to circumstances beyond control. Certain kinds of degradation exist by natural law; other kinds are imposed by limitations of the evolving technology of image reproduction. Photographers would do well to learn which kinds can be controlled and which kinds can be reduced only so far. Those who do not are apt to waste time and effort trying to achieve the unattainable.

Chapter 9

Light and Exposure
How Light Influences Image Clarity

Photographers can exploit certain properties of light—its quality, direction, brightness, and wavelength—to enhance image clarity and increase the resolution of detail. On the other hand, light can reduce image information during exposure and by way of diffraction. The better you understand how light affects the clarity of reproduction, the better your ability to optimize detail and sharpness in your photographs.

Quality of Light

The quality of light refers to the collimation of the rays in a beam of light, not to ethereal quality. This property of light is judged simply by the quality or sharpness of shadows. The highly collimated rays that create sharp shadows are said to be directional or hard; uncollimated rays are said to be diffuse or soft.

Directional Lighting

Directional lighting emphasizes surface textures and fine details. Such lighting creates well-defined shadows that echo miniature surface shapes. These shadows are not superfluous but are vital to textural clarity. They are the visual clues that prove the image has been accurately reproduced. Even when they are small and are seen superficially, they reveal surface characteristics more clearly than would otherwise be possible.

When light from a directional source reflects specularly from textured surfaces toward the lens, bright highlights are created. Specular highlights are those reflections of the light source, from shiny objects, for example, that are nearly as intense as the light source itself. These highlights can add as much to clarity as shadows do. (See Plate 2.) When small, crisp highlights appear, they confirm that the detail and sharpness of an image are good. They also cause the image literally to sparkle.

Directional lighting comes from distant point sources or from focused light sources. A *point source* is a theoretical device approximated by a source of very small size or by a light located a great distance from the subject. Generally, for a lamp to have a point-source effect, it must simply be small in relation to the size of the subject. A certain lamp may resemble a point source at ordinary subject distances but may resemble a broad source in close-up work. The sun is a directional light source; by the time its rays reach the earth they are almost perfectly parallel. Because of its distance, it is nearly a point source. Although solar shadows have narrow fringes of softness that prove the sun is not a true point source, it is difficult in practice to find more directional lighting than that from the sun.

Focused light sources, such as photographic and theatrical spotlights, produce rays that are nearly parallel. Spotlights have intense light sources positioned between a polished parabolic reflector and a Fresnel lens. The reflector collects the rays and directs them toward the subject. The lens focuses the rays to prevent them from diverging along the way.

Diffuse Lighting

Diffuse light sources are large with respect to the subject and emit light as though many separate point sources were deployed over the radiating surface. Textural shadows are neutralized under diffuse lighting as light arrives at the subject from virtually all directions. Unless textures are coarse, their visibility can be weakened considerably under diffuse lighting. (See Plate 3.)

Diffuse lighting also weakens the perception of edge sharpness. The quality of light has no direct bearing on the reproduction of edges, but as you have seen before, the hard shadows created by directional lighting give important visual clues about sharpness. Diffuse lighting introduces soft shadows that eliminate these clues. In diffuse lighting, edges may appear softer than they really are.

Direction of Light

The direction of light refers to the angle at which light strikes the subject with respect to the lens axis. The direction of light determines how well the form and contours of an object are revealed and, as collimated light does, influences the clarity of surface textures by the kind of shadows it creates.

Angular Lighting

Angular lighting refers to lighting that strikes the subject at a substantial angle to the lens axis, on the order of 30 to 60 degrees. Lighting from such angles creates the kind of shadows that enhance clarity while giving the image a pleasing, natural appearance. Lighting from even steeper angles, from 70 to 90 degrees, known as *crosslighting,* gives greater emphasis to surface textures and, up to a point, renders textures quite dramatically. But at extremely oblique lighting angles, shadows dominate the composition and the image may lose detail or appear unnatural.

Dominant backlighting is far more useful than many photographers realize. It defines the form, shape, and texture of a subject, often better than frontal, angular lighting does. It separates a subject clearly from its background by adding a rim of light to the subject's outline. It casts a foreground shadow that repeats and clarifies the shape of the subject. And where it skims the surface it creates specular highlights that improve textural clarity and that enhance visual interest by adding "sparkle" to the image. The beauty and effectiveness of backlighting is well understood by expert lighting technicians but often goes unappreciated by inexperienced photographers.

Dominant Backlighting

Axial lighting is created by a light source positioned frontally in line with the lens axis. It is very flat—that is, it makes contoured surfaces seem flatter than they are by the even, shadowless illumination it gives. Since the shadows it creates are not detected at the film plane, textural contrast is minimized in axial lighting. In fact, it renders textural features so blandly that it is sometimes used deliberately to suppress the visibility of small features, blemishes, and surface flaws.

Axial Lighting

Technically, axial lighting is achieved by placing the primary lamp directly on the lens axis. If this practice were followed strictly, however, the lamp would appear in the picture or would be blocked by the camera. Ringlights, circular lamps that surround the lens, give shadowless illumination which closely simulates that of precise axial lighting; but the source of light, though near the axis, is not precisely on it. Precise axial lighting can be achieved with a half-silvered or two-way mirror. The mirror is placed between the lens and subject at a 45-degree angle. Light is reflected by the mirror onto the subject from a lamp placed at a right angle to the lens axis. See Figure 9.1.

Precise Axial Lighting

Narrow lighting angles from 10 to 20 degrees or less, such as are achieved with many camera-mounted flash units, often produce an undesirable lighting effect. Such angles produce near-axial lighting that not only weakens textural details, but puts a disturbing peripheral shadow close to the subject that follows the subject's outline. Such a shadow is as likely to obscure the subject's shape as to enhance it.

Near-Axial Lighting

Diffraction is often associated with lenses. Despite the influence lens aperture has on it, however, diffraction is a natural phenomenon

Diffraction

FIGURE 9.1
Precise axial light

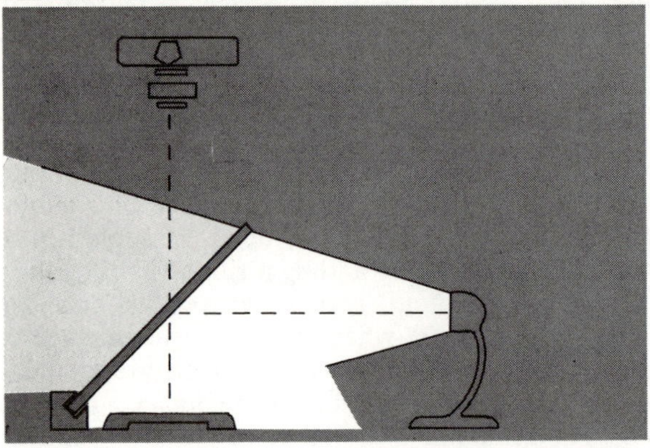

A beam of light can be placed precisely on the lens axis with a beam splitter, which can be a sheet of glass or a half-silvered mirror. The mirror is placed between the camera and subject at a 45-degree angle. Light is directed at the beam splitter at a right angle to the lens axis.

arising from the properties of light itself and cannot be eliminated by making better lenses.

The consequences of diffraction can be seen directly by examining a greatly magnified image of a point-sized object. Indeed, diffraction was first observed through astronomical telescopes in images of distant point-sized stars. Even in today's highly corrected telescope systems, images of such stars cannot be rendered as absolute points as one might expect. They are rendered instead as patterns of light consisting of a bright central disk surrounded by several concentric rings of dimmer light. Surprisingly, reducing the size of the admitting aperture of the lens can have the effect of increasing rather than decreasing the dimensions of a point-sized image recorded on film. Also, at a given aperture size the central disk can become just so small. A time comes when a point-like subject gets smaller, but its image does not. This diffraction phenomenon was described in 1830 by the astronomer G. B. Airy. It is accounted for by the wave nature of light.

Diffraction occurs because of what happens to a wavefront when light passes an obstruction. First consider what happens in an unobstructed wavefront. Each point on an open wavefront acts as an independent radiator that generates its own secondary wave. If these secondary waves were to continue radiating independently, they would all diverge and the main wavefront would disperse; but interior waves interact with others on either side. By a process of destructive interference, the independent radiations in an open, unobstructed wavefront precisely cancel and reinforce each other in a way that keeps the wavefront intact.

When the wavefront passes an obstruction, however, it emerges with an edge at which secondary radiations are not entirely eliminated. Beyond the obstruction, secondary waves near the obstructed edge radiate in all directions, but unlike before, there are no neighboring waves on the obstructed side to interfere. Light near the edge is diffracted and radiates independently of the main wavefront. Some of the light spreads into and illuminates the shadow near the edge; some spreads into and adds to the illumination in the main wave. See Figure 9.2.

When light passes through an aperture of finite size, such as a pinhole or the diaphragm of a lens, that part of the light passing near the circular edge of the aperture opening is diffracted. See Figure 9.3. The diffracted rays cannot be made to converge to a precise point by optical devices. No matter how small the pinhole and no matter how perfect the optical system, the independent radiations arising from this diffraction component will always prevent image points from being immeasurably small.

For the photographer, however, the situation is not all grim. If the aperture is large, the central area of the beam will contain a healthy, undiffracted component. In a well-corrected optical system this undiffracted portion of the wave can be focused to a virtually perfect point and is capable of creating images of great clarity.

The degree to which diffraction degrades a photographic image depends mainly on the ratio of diffracted light to undiffracted light in an image point. If the clear, undiffracted component is dominant, its influence will predominate over exposure and the diffracted part will have little practical effect on the image.

The ratio of diffracted to undiffracted light varies with the size of the aperture. Diffracted light near the edge varies in proportion to the circumference of the beam, a linear factor of the radius ($C = 2\pi r$), while undiffracted light in the central portion of the beam varies in proportion to the area of the beam, a square factor of the radius ($A = \pi r^2$). Image points formed by a large aperture contain large amounts of diffracted light passing near the edge, but the

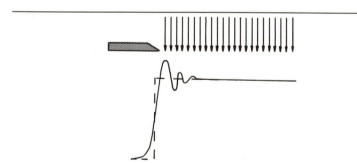

FIGURE 9.2
Edge diffraction

As a wavefront passes an obstruction, secondary waves formed at the newly created edge radiate into areas shaded by the obstruction. These secondary waves interact with wavelets in the main wave—reinforcing some, canceling others—to create bands of light of alternating intensity.

percentage of unaffected light passing through the center is proportionately larger. Both diffracted light and undiffracted light increase as the aperture increases, but undiffracted light increases faster. See Figure 9.4.

When light passes through a small aperture, diffraction predominates. In this case the diffracted light becomes bright enough in relation to the brightness of the main wave to be seen and recorded as part of the image. Through very small apertures virtually all light finding its way past the diaphragm is diffracted, effectively overpowering image light to increase point size in the recorded image.

The aperture diffraction pattern represents the smallest image point that can ever be rendered by a lens or a pinhole at a given aperture size. In effect, though, it is the disk of light at the center of the circular diffraction pattern, the Airy disk, that is significant. The Airy disk is 12 times brighter than the brightest ring that surrounds it. It is therefore the only part of the pattern that is photographically important. The radius of this central disk is given by

$$r = 1.22\lambda \frac{F}{a}$$

where λ is the wavelength of light, F is the focal length of the lens, and a is the relative aperture of the lens. Substituting 0.0005 millimeters into the equation for λ as a representative value for the wavelength of visible light and substituting the f/number N for F/a one arrives at

$$r = 0.00061 \ N.$$

FIGURE 9.3
*Aperture
diffraction*

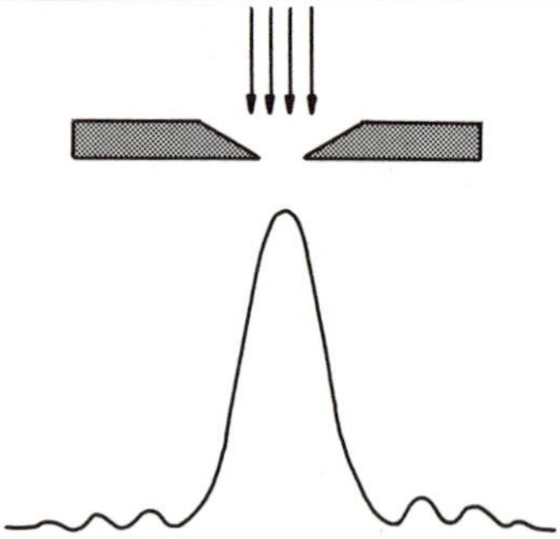

Diffraction occurs all around the edge of a circular aperture. The resulting optical pattern consists of a bright central disk surrounded by rings of dimmer illumination. The central disk is called the Airy disk after the astronomer G. B. Airy, who first described it.

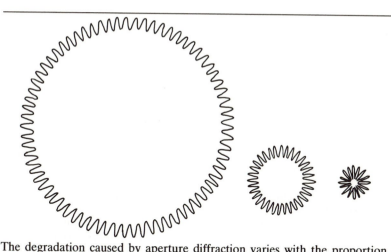

FIGURE 9.4
*Diffraction at large
and small apertures*

The degradation caused by aperture diffraction varies with the proportion of diffracted light around the edge of the image beam to undiffracted light in the center. Although there is much diffracted light around the circumference of a large beam, there is far more undiffracted light in the central area of the beam. The unaffected central beam, which can be focused with incredible sharpness and which is proportionally brighter, controls exposure; the diffracted light has little practical effect. At small apertures diffracted light at the edge of the beam is so large a part of the total light that it dominates in the exposure and becomes a visible part of image points.

Since the f/number N is now the only variable on the right side of the equation and since it must always be of finite size, the diffraction-spread component represented by r must always be of finite size. This equation defines a fundamental physical limit. It shows how, in images made by visible light, there will be an unavoidable kind of degradation controlled by the f/number of the lens and by nothing else. Such degradation cannot be eliminated by technique or technology. Indeed, image points cannot be made smaller than the diffraction limit by any photographic procedure. In lenses where diffraction is the only factor that limits resolving power, the lens is said to be *diffraction limited*. A diffraction-limited lens is as nearly perfect as a lens can be. The resolving power R of such a lens can be estimated using the relationship

$$R = \frac{1}{r}$$

where r is the radius of the Airy disk.

To the extent that the wavelength of imaging light affects diffraction, it affects resolving power. Since wavelength determines the color of

Wavelength of Light

light, the spread function of an optical system varies in theory with the color of imaging light. In fact, short wavelengths, those at or beyond the ultraviolet end of the spectrum, produce smaller diffraction components than do longer wavelengths at the infrared end of the spectrum. Other things being equal, images made in ultraviolet radiation will resolve finer detail than images made in longer wavelengths.

The improved resolving power of ultraviolet radiation, however, has few practical applications in ordinary photography because glass lenses absorb radiation having wavelengths shorter than about 3500 angstroms. Although silver-halide films will respond to these and even shorter wavelengths, the improved resolving power of ultraviolet radiation cannot be exploited using ordinary lenses. Ultraviolet radiation is therefore of slight value for increasing resolving power in pictorial photography, though it has found applications in other areas, such as ultraviolet microscopy, photomicrography, micrographics, and integrated circuit technology.

Level of Light

The level of light, because it determines the choice of aperture and shutter speed, affects depth of field and blurring from image motion. Dim light is a major obstacle to obtaining high-resolution images. It requires that either exposure time be long, risking undue image motion, or that a large aperture setting be used, reducing depth of field and increasing point spreading from aberrations. The odds are,

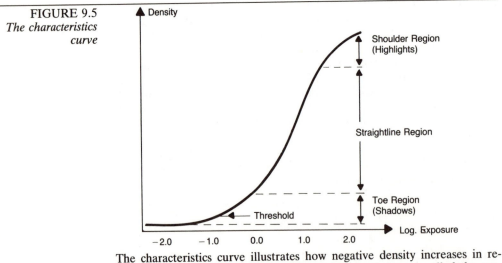

FIGURE 9.5
The characteristics curve

The characteristics curve illustrates how negative density increases in response to increases in exposure. The region on the curve called the toe, which corresponds to exposure just above threshold, is associated with subject shadows. The straightline region of the curve is associated with middle values, the shoulder with highlights.

in either case, that resolving power will be lower at low light levels than at high light levels.

Errors in exposure can reduce image clarity by their influence on image contrast and by their influence on emulsion turbidity.

Exposure and Image Clarity

Exposure and Contrast

A film's characteristics curve shows how exposure can change the density and contrast of an image. See Figure 9.5. Exposure is indicated on the characteristics curve along the horizontal axis in units called *meter-candle-seconds* (MCS). A meter-candle-second is the exposure received in one second by a film held one meter from a light source of one standard candle intensity. Exposure is usually plotted in logarithmic units. For example, the number -2.0 on the horizontal axis of the curve indicates 0.01 MCS and 1.0 indicates 10 MCS.

The characteristics curve shows that two contrasting subject tones will record in the negative with a smaller difference in density and with lower contrast if they are exposed on the toe or shoulder of the characteristics curve than if they are exposed on the straightline portion. See Figure 9.6. This is because density increases less in the toe and shoulder regions than in the straightline region of the curve in response to a fixed increase in exposure. This effect is aggravated by overexposure or underexposure. Generally, contrast and clarity are degraded most in shadows upon underexposure and most in highlights upon overexposure.

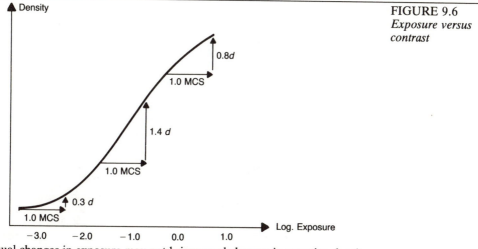

FIGURE 9.6
Exposure versus contrast

Equal changes in exposure may not bring equal changes in negative density. Density increases less on the toe or the shoulder of the curve than it does on the straightline portion for a given increase in exposure.

FIGURE 9.7
Turbidity

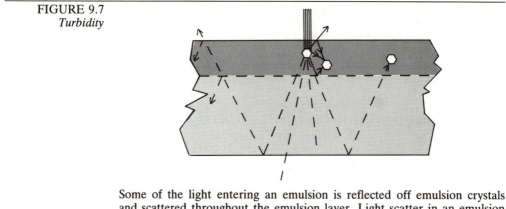

Some of the light entering an emulsion is reflected off emulsion crystals and scattered throughout the emulsion layer. Light scatter in an emulsion is a measure of its turbidity.

Exposure and the Spread Function

Overexposure reduces clarity in a second, more significant way: it worsens turbidity. See Figure 9.7. It increases light scatter in the emulsion and enlarges the emulsion spread function. Overexposure is especially harmful in thick-emulsion films because of the greater depth at which scattering can occur.

The damage done to image clarity by overexposing a film will depend on the magnitude of the error and on the properties of the

FIGURE 9.8
Low and moderate exposure

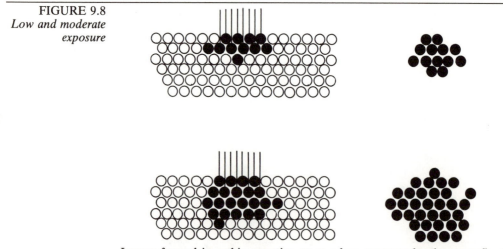

Images formed in a thin negative or at a low exposure level are confined to the surface of the emulsion. As exposure increases, the image forms deeper within the emulsion layer and expands laterally because of emulsion turbidity.

film. See Figure 9.8. Exposing a coarse-grain, wide-latitude film by one f-stop or less above its optimally correct exposure may not cause enough damage for it to show up above the film's massive spread function. Exposure errors greater than two or three f-stops above optimum, however, may scatter so much light in the emulsion layer that the deterioration in detail and sharpness can be seen at relatively low enlargement ratios. At the far extreme, there is no limit to the damage that can be done by an exposure so massive as to cause halation. See Figure 9.9.

Thin-emulsion films on the other hand, with their low tolerance for overexposure, can be ruined by an exposure error as small as one f-stop above optimum (perhaps less). The problem here is that the maximum silver density of a thin-emulsion film is easily reached. The first symptom of overexposure in these films will more likely be blockage of tonal values in highlights rather than light scatter.

Overexposure that results in halation is especially harmful. *Halation* is the consequence of extreme overexposure such as occurs when intense light, as from a light source, strikes the film. An exposure great enough to cause halation must be so great that part of the light passes completely through the emulsion layer and support base of a film. The light is then reflected back by the rear surface of the

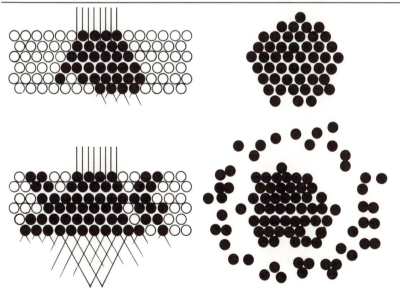

FIGURE 9.9
*Overexposure and
halation*

Massive exposure creates images of maximum density that penetrate to the deepest parts of the emulsion layer. Severe overexposure can cause halation—exposure must be great enough for light to pass through the emulsion layer, reflect from the film base, and expose the emulsion layer a second time. Halation often occurs in the images of light sources or intense specular reflections. There is no limit to the damage that can be done by halation.

support base so that it again exposes the emulsion layer. Halation causes severe light spreading and a general fogging in the area surrounding bright light sources in the image. In some instances the halation pattern takes the shape of a ring of light (like a halo) surrounding the source.

Chapter 10

Contrast and Gradation
The Importance of Tonal Distinctions

Not all photographers understand how damaging the loss of contrast can be to resolving power, but contrast is clearly essential to the visibility of fine detail. It was shown in Chapter 3, for example, that contrast is the controlling parameter in the modulation-transfer function and that it is a key element in the acutance formula. It was shown in Chapter 7 that when image points spread and overlap adjacent points, it is the loss of contrast that determines whether these points will be resolved. In fact, every method so far examined for measuring image clarity demonstrates that the limit of resolution is reached when contrast is reduced below the visual contrast threshold.

To appreciate better the relationship between contrast and resolution, one must make a distinction between general and local contrast. General contrast is sometimes associated with image gradation—the range of intermediate shades of gray reproduced in a photograph. Local contrast, contrast within the microstructure of an image, governs the resolution of detail and textural clarity. It is evidenced by the presence of tiny shadows and highlights in the smallest visible surface structures. These tiny highlights and shadows, by their size and sharpness, reveal the substance and character of a surface. They help the viewer visually differentiate matte surfaces from glossy ones and smooth surfaces from porous or fibrous ones. When local contrast is weakened or diluted, image details may simply merge in tone with the background and disappear.

Several important ideas about photographic contrast are examined in this chapter: the connection between contrast and gradation is clarified and a method is examined for determining when contrast and gradation are present in an image in ideal proportions. The chapter ends with an examination of common kinds of contrast degradation, such as flare and tonal mergers. First, some of the terms used to describe contrast are discussed.

Photographic Contrast

Contrast refers to the extremes of tone, density, or luminance in an image. It is sometimes stated in terms of a *range,* that is, the arithmetic separation between the extremes. Range is generally used in reference to densities and log exposure, both of which involve logarithms. Contrast may also be stated as the *ratio* of the greatest to the smallest extreme. Ratio is usually associated with luminance and illuminance, nonlogarithmic numbers. If the difference in luminance extends from a low of 5 units to a high of 100 units, the contrast ratio will be 100 to 5, which is usually reduced and written as 20 to 1 or 20:1. The contrast *scale* is a single number found by taking the quotient of the greatest to the least luminance. In the example above, the scale is simply 20.

Print Contrast

The greatest contrast attainable in a print is limited by the reflectance ratio of the print paper, that is, the ratio of reflectance of the white paper base to the reflectance of the maximum silver density of the emulsion. A print paper with a matte or lustre finish may have a reflectance ratio of 50 to 1 or less. Certain glossy black-and-white papers have a reflectance ratio of about 80 to 1.

Keep in mind the difference between a print paper's reflectance ratio and its *contrast grade*. The reflectance ratio of various classes of paper, glossy or matte, is somewhat fixed within each class. A glossy variable contrast paper, for example, cannot exceed its maximum reflectance ratio, around 80 to 1, regardless of the contrast filter used with it. The full range may or may not be brought out in a particular print, but this ratio cannot be exceeded. Contrast filters, or the paper's grade, are related to the exposure scale of the paper, the ratio of the exposure needed in shadows and the exposure needed in highlights to reach the maximum print reflectance range. This hinges, in turn, on how thin or contrasty the negative is.

Subject Luminance Scale

Variations in subject reflectance and variations in incident illumination from one scene to the next complicate the matter of subject contrast and make mastery over exposure and tone control all the more difficult.

Reflectance Ratio.   Reflectance is a measure of the amount of light reflected from a surface as a percentage of the light striking it. Reflectance ratio R is the ratio of the highest reflectance in a scene r_{max}, that associated with the lightest or whitest object, to that of the lowest reflectance r_{min} associated with the darkest or blackest object:

$$R = \frac{r_{\max}}{r_{\min}}.$$

The most reflective or whitest natural objects reflect around 96 percent of the light that strikes them. The least reflective or blackest objects reflect about 2 percent, absorbing the remaining 98 percent. The greatest reflectance ratio one is likely to encounter naturally is therefore about 48 to 1. But one should generally expect to find a lower ratio; both extremes are rarely encountered at once by chance.

Lighting Ratio.   Lighting ratio indicates the ratio of luminance between primary and secondary light sources:

$$L = \frac{P}{S}$$

where L is the lighting ratio, P is the luminance of the primary light as measured at the subject location, and S is the luminance of the secondary light, also measured at the subject location. In natural daylight the lighting ratio can range from 1 to 1 on an overcast day to 10 to 1 on a clear, sunny day. This ratio does not necessarily measure lighting contrast. It does not, like the contrast formula below, account for the possibility that the primary and secondary lights may overlap in their coverage of the subject.

Lighting Contrast.   Lighting contrast indicates the proportion of light to shade in a scene. Under conditions of controlled lighting, such as in a studio, one can compute lighting contrast C in terms of the primary and secondary lights:

$$C = \frac{P + S}{S} = L + 1.$$

Outside the studio, tertiary shadows and illumination caused by reflectors and shades can alter lighting contrast to such an extent that this formula may not correctly predict the full contrast scale of a natural sunlit scene. The formula accounts for the overlapping of the primary and secondary lights, but it does not account for the possibility that part of a subject may be shaded from both the primary and secondary sources.

Lighting contrast is more correctly determined by directly measuring the level of light in various parts of a scene as modified by reflectors and shades:

$$C = \frac{i_{\max}}{i_{\min}}$$

where i_{max} is the illumination falling in the most luminous area of a scene, as in bright sunlight, and i_{min} is that falling in the dimmest parts of the scene. So although the natural lighting ratio, the ratio of sunlight to open shade, may be about 10 to 1 or less, contrast in natural lighting can be much higher. The presence of deep shadows, in which neither sunlight nor skylight reach the subject, can raise lighting contrast to 100 to 1 or more.

Luminance Scale. While reflectance is influenced by properties of the subject and lighting contrast is influenced by the nature of the lighting, luminance or brightness is influenced by both. Luminance measures the absolute quantity of light given off by an object as affected by both the level of incident illumination and the reflectance of the object's surfaces. The most luminous area in a scene is found where the brightest light strikes the most reflective surface. A white object in bright sunlight, for example, has a relatively high brightness. A black object shaded from both the primary and the secondary light sources has a relatively low luminance. The relationship is expressed by:

$$L = RC$$

where L is the luminance scale, R is the reflectance scale, and C is the lighting contrast. If the subject's reflectance ratio is 30 to 1 and the lighting contrast is 4 to 1, the luminance scale will be the product of the two, or 120.

Thus there is a wide range over which subject luminance can vary. In even, shadowless lighting, the luminance scale will depend entirely on the reflectance ratio of the subject, which will always be below 48 to 1 or so. On a sunny day the luminance scale can be far greater. Variations occur from subject to subject, from hour to hour, from day to day, and from season to season. In one study of this variation, luminance measurements were taken at 150 different outdoor scenes over a period of one year. The luminance scale varied from a low of 27 to a high of 750. References to the luminance scale of a typical or an average daylight scene usually assume a scale of about 160.

Gradation

Gradation refers to the number of separate, recognizable image tones or shades of gray in an image. Gradation and contrast are related concepts in that gradation is influenced by the rate of change of contrast. Gradation and contrast are otherwise unique. Nevertheless, contrast is sometimes referred to when gradation is meant. A print that contains stark white, deep black, and no intermediate gray tones is commonly called a high-contrast print. Yet a print containing a full scale of gray tones, including stark white and deep black, though it differs in gradation, is equal in contrast to the first. Contrast is not necessarily diminished by an increase in gradation.

Poor gradation, like poor resolving power, is symptomatic of a photographic system's inadequate information capacity; gradation consumes information capacity just as resolving power does. Theoretically, one of these properties can be improved upon only at the expense of the other. You will recall that gradation is alluded to mathematically in the formula introduced in Chapter 3 that defines information capacity:

$$C = n(\log_2 d)$$

where n is the number of elemental areas in an image and d is the number of unique image tones or density levels. This formula confirms that, in systems of equal information capacity, one system can be structured to contain many elemental areas (high resolving power) and the other to contain many density levels (wide tonal scale). Improving one quality without sacrificing the other requires that information capacity be increased.

The formula also leads one to suspect that image tones are elements of information, and indeed they are. Gradation provides visual information that improves one's recognition of and satisfaction with an image.

Ideal Contrast and Gradation

Ideal Lighting Contrast

Although the visibility of fine detail improves as contrast increases, there is a practical limit in photography to how far one can take this improvement. One cannot increase lighting contrast indefinitely without weakening the gradation of tones in the printed image. Extreme lighting contrast creates stark highlights and deep shadows in prints that reduce tonality and destroy information. This happens because photographic emulsions cannot accommodate such extremes in contrast, or can only do so with curtailed development. The range of acceptable contrast ratios, ratios which retain both clarity of fine detail and variation of image tonality, is somewhat narrow.

Natural Lighting Contrast.   One can develop an intuitive understanding of lighting contrast by analyzing the range of contrast commonly found in natural daylight. One must first understand the distinction between sunlight, skylight, and daylight. *Sunlight* refers to the direct rays of the sun, *skylight* is light reflected within the atmosphere, and *daylight* is any combination of natural sun and shade. Do not underestimate the importance of the sky as a source of light. Indeed, skylight is the reason objects shaded from direct sunlight can be seen at all. Without it, shadows would be pitch black and devoid of detail.

From an exposure table you will find that a subject in open shade needs about three f-stops more exposure than one in bright sunlight. This means that on a typical clear day, skylight is about one-eighth as bright as direct sunlight and the lighting ratio is about 8 to 1.

The precise ratio varies, but one can expect the lighting ratio in natural sunlight to vary over a range from 5 and 10 to 1.

Photographs made in direct, unmodified sunlight tend to have harsh tonality that is especially obvious and unbecoming in portraiture. When areas in full sunlight are rendered with normal tonalities, shaded areas contain large masses of dark tones with little detail. A lighting ratio of 8 to 1 in studio photographs will create the same kind of dense shadows and harsh highlights as in photographs made by sunlight.

A suitable lighting ratio for most kinds of photography usually falls somewhere between the extremes found in natural daylight. Lighting ratios of about 2 or 3 to 1 such as may occur on a bright, hazy day or the even lighting of an overcast day are generally more acceptable in studio and portrait photography. Many commercial photographers routinely use a 3-to-1 ratio, trying never to exceed a ratio of 4 or 5 to 1. If higher lighting ratios are used, it becomes difficult to hold detail simultaneously in the shadows, midtones, and highlights of prints.

Ideal Print Tonality

The photograph is the final proof of the soundness of your technique. So it is in the print more than in the negative that contrast and gradation matter.

Studies indicate that viewers prefer prints that use the full reflection density of the print paper and display a wide range of tonal values. In other words, the preferred print contains pure white, solid black, and many shades of gray. Consequently, these characteristics are the ones sought by careful printers. The presence of pure white and deep black means that print contrast is at its maximum. The presence of many shades of gray means that print gradation is also at maximum. A fine print is therefore one in which contrast and gradation are concurrently maximized.

Ideal Negative Contrast

The tonal characteristics of the negative, an intermediate step in image formation, are of less concern than the contrast and tonality of prints made from it. Ideal negative contrast is simply that which produces ideal print tonality as discussed above. Thus the parameters used to judge the tonal quality of negatives, shadow detail and highlight detail, are more appropriately determined jointly from the properties of a negative and from the best print made from it.

It takes a special kind of negative to create an ideal print in which contrast and gradation are simultaneously maximized. Excessively thin negatives print with both low contrast and low gradation. Dense, contrasty negatives create strong print contrast but harsh gradation. Print gradation increases initially as the negative's contrast increases. It is best when the negative's density range matches or slightly exceeds the print paper's exposure scale, but declines with further increases in negative contrast.

At the print exposure that gives solid black in deep shadows, the negative's highlight density should be such that specular highlights print as pure white and diffuse highlights print with a detectable density just darker than the maximum white of the paper base. Specular highlights, as mentioned previously, are bright, direct reflections of the light source. Diffuse highlights are bright tones that contain detail and texture. Getting texture and brilliance in diffuse highlights is best ensured by tailoring the density range of the negative to fit the printing characteristics of the paper and the enlarger used. The important consideration here is that the optimum film development time be used. A systematic way to maximize print contrast and gradation concurrently is discussed in Chapter 17.

Ideal Gradation

Gradation, like resolving power, can deteriorate considerably before an image is judged unacceptable. Images containing two tones, the high-contrast, low-gradation kind, are harsh and are not always visually appealing. Even so, one can usually recognize major features in such images and can often determine something about the subject's form and substance.

The presence of one additional tone, as in a three-toned posterization, increases image appeal and realism considerably. (See Plate 4.) However, three-toned images rarely suggest the presence of a full tonal scale. Images containing as few as five tones are often perceived as having complete gradation.

The presence of more than five image tones greatly increases one's satisfaction with image tonality. Yet although image information increases as gradation improves, the eye cannot distinguish an infinite number of image tones. Depending on ambient conditions, luminance differences must be greater than 1 to 3 percent to be detected. This means that one can see 60 shades of gray at most in good conditions and fewer in average or poor conditions. Thus any image that traverses the tonal scale from black to white in 60 or more increments of density contains about as much gradation and tonality as can be detected by the eye; such an image cannot be distinguished visually from one of continuous gradation.

Degradation of Contrast

Low contrast within the microstructure of an image reduces the clarity of detail. Excessive contrast, either at the subject or in the negative, reduces the gradation of tones. Ideally one should seek to preserve contrast at an optimum level by moderating lighting contrast on the subject, developing the negative to its proper contrast, and avoiding the inadvertent degradation of contrast at any point in the imaging process. Be wary of flare, curtailed development, excessive fill lighting, and tonal mergers.

Flare

Flare consists of nonimage light that veils the image and reduces its contrast in the process. For details about how it does so, see Chapter 14.

Curtailed Development

Often photographic contrast must be reduced out of necessity so that the luminance values of the subject can be made to fit within the narrow reflectance range of an available photographic paper. In doing so, image tones are compressed, usually by curtailing or reducing the degree of film development. Such a compression is essential to prevent the loss of detail in shadows or highlights that would occur if the negative's density range were to exceed the paper's exposure scale. Nevertheless, this compression of tones somewhat reduces the contrast and visibility of fine detail. The high-resolution photographer will be cautious about curtailing development unnecessarily if a better way can be found to control subject contrast.

Excessive Fill Lighting

In stark, contrasty lighting it is a good idea to reduce general lighting contrast by using secondary fill lights to illuminate shaded areas. Fill lighting is often vital to the visibility of objects in shadows, but it can be taken too far. If you weaken contrast in the microstructure of an image to promote general tonality, resolution will be lost. Much depends on your fill-lighting technique. You will see in Chapter 21 that textural contrast can be controlled independently of general contrast by thoughtful lighting techniques.

Tonal Mergers

Tonal mergers, a greater problem in black-and-white than in color photography, occur when the tonal values of adjacent image features are so similar that one cannot tell where one ends and the other begins. They occur, for example, when shaded parts of a subject appear next to a shaded background. When adjacent tones merge, the viewer is denied complete information about the subject. Its form and outline are concealed. Master photographers are usually quite conscious of tonal mergers from the moment they compose an image in the viewfinder. For example, much of the feeling of sharpness in the pictures of masters like Ansel Adams comes from their having carefully avoided letting matching tones meet. Techniques designed to detect and eliminate tonal mergers are discussed in Chapter 19.

Chapter 11

Photographic Films
Limitations of the Silver-Halide Process

The ultimate acutance, gradation, and resolving power of a film are fixed by three properties of its emulsion layer: average crystal size, emulsion thickness, and range of crystal size. An understanding of these properties will help you to evaluate the usefulness of various films for high-resolution jobs. This chapter begins by showing how photographic emulsions are prepared, how they respond to light, and how their physical properties affect a film's imaging characteristics.

Emulsion Preparation

A photographic emulsion consists of microscopic crystals of one of the silver halides, often silver bromide, suspended in gelatin. The first step in making an emulsion is called *precipitation*. In this process light-sensitive crystals, usually silver-bromide crystals, are precipitated by slowly mixing silver nitrate with potassium bromide. If these compounds were mixed in water the crystals formed would not only be coarse and granular, but would sink to the bottom of the vessel containing them. Instead they are mixed in a gelatin solution that suspends the crystals and keeps them randomly dispersed. Gelatin also slows the rate of crystal growth, allowing it to be better controlled.

When crystals are first formed they are small and react slowly to the action of light. During the second step of emulsion preparation, *physical ripening,* crystals grow in both size and sensitivity. In this step the emulsion is heated and allowed to sit for an hour or two. As the crystals ripen, opposing forces encourage them on the one hand to dissolve and on the other hand to crystallize and grow. At these sizes the tendency is for smaller crystals to dissolve faster than they crystallize and for larger crystals to grow faster than they dissolve. The larger crystals thus increase in size at the expense of smaller ones in a process known as *Ostwald ripening*.

Crystals also grow by another process. When the distance between crystals approaches a little more than the diameter of the crystal, they tend to gravitate toward one another and associate into groups. Eventually, if the ripening process continues, crystals will physically touch and coalesce. This phenomenon is known as *coalescent ripening*.

Following physical ripening the emulsion is either shredded and washed or chemically treated to remove and neutralize unwanted by-products, then allowed to ripen further. During this step, called *after-ripening*, the emulsion is reheated while various chemicals, including hardeners and spectral sensitizers, are added. All that remains beyond this step is to coat the emulsion onto a support base for use. Photographic emulsions are coated onto support bases of plastic, glass, or paper in the manufacture of photographic films, plates, and papers.

Light-Sensitive Crystals

The silver halides commonly used in photography are silver bromide, silver chloride, and silver iodide. Of the three, silver bromide is the fastest, the most sensitive to the action of light, and hence the most widely used. It is the primary halide in virtually all commercial emulsions made for films and is the primary halide in most emulsions made for photographic papers.

Silver chloride is the next fastest, followed by silver iodide. Silver chloride and silver iodide are never used alone in emulsions intended for commercial films. Silver chloride is sometimes combined with silver bromide in emulsions intended for photographic papers. It is occasionally used alone in making certain contact printing papers for which a massive print exposure is feasible. Silver iodide, when used in an emulsion at all, is generally combined in small amounts with another silver halide; then its main function is to alter the lattice structure of the halide crystal.

Light and the Silver-Halide Crystal

Sensitivity Specks

Surprisingly, chemically pure and structurally perfect silver-halide crystals react slowly to light. Contaminated crystals, particularly those containing certain sulfur compounds, react faster than crystals of high chemical purity. Crystals with defective lattice structures are more light sensitive than those with orderly structures. It is to create such lattice defects that silver-bromide emulsions are doped by the introduction of a trace of silver iodide. By altering the regular crystalline lattice pattern, silver iodide improves photographic sensitivity.

The structural defects and chemical impurities that improve the sensitivity of silver-halide crystals to light are rightly called *sensitivity specks*. Sensitivity specks seem to operate as traps or collection sites for free electrons. When light strikes a silver-halide crystal and is absorbed, an electron is released. If the crystal has a sensitivity speck or electron trap, this electron finds its way to it, giving the speck a negative charge. The negatively charged sensitivity speck then attracts positively charged silver ions which move freely through the crystal. When a silver ion combines with a free electron at the sensitivity speck, an elemental atom of silver is formed which has the effect of enlarging the speck and making it an even more effective electron trap. If more electrons are released immediately thereafter, the process repeats itself and is likely to do so at the enlarged site.

A single atom of silver at a sensitivity speck is unstable, surviving for perhaps a fraction of a second before rehalogenating, that is, reverting to a silver ion by shedding the electron and recombining with a bromide ion to form a new molecule of silver bromide. Two atoms of silver make the sensitivity speck stable for a day or so. Exposure is thus cumulative, provided it is great enough to generate additional silver atoms before the first rehalogenates. This rehalogenation of silver atoms at the sensitivity speck is the basis for the low-level failure of the *reciprocity law* discussed in Chapter 16.

Latent Image Centers

A *developable* sensitivity speck is called a *latent image center*. As few as three atoms of elemental silver at a sensitivity speck seem sufficient to render the speck both stable and developable. At this size the speck of silver at the trap is apparently large enough to provide a chemical point of attack allowing development to be induced. The larger a latent image center becomes during exposure, the sooner development of the crystal starts. Crystals in which exposure is insufficient to create a latent image center either do not develop, develop with difficulty, or lose their silver speck by way of rehalogenation.

Emulsion Crystal Size

The silver-halide crystals in photographic emulsions range in size from as small as 0.01 to 0.05 microns in micro-fine-grain Lippmann-type emulsions to as large as 2 to 4 microns in super-coarse-grain X ray films. See Figure 11.1. The size of emulsion crystals determines three important properties of a film: resolving power, sensitivity to light, and graininess.

Resolving Power

The resolving power of a film is limited by the size of its crystals in the sense that image features smaller than the smallest emulsion crystals cannot be separately recorded. Yet it is no simple matter to express this relationship mathematically. Formulas have been advanced but have proven valid only for the specific films used to derive them. The relationship is complicated by other factors. Turbidity and exposure, for example, can so enlarge the emulsion-spread function that they place the principal limit on resolving power.

Emulsion Sensitivity

If two emulsions differ from one another only in the average size of their crystals, the emulsion having the larger crystals will respond more readily to scenes of low brightness. The reason is that the threshold exposure of a crystal, the minimum exposure needed to make the crystal developable, is about three photons of light. This minimum exposure is thought to be independent of crystal size. Since crystals with large projective surfaces will be struck more often than smaller crystals when both are exposed to identical illumination, large crystals absorb more light and are more likely to form stable latent images.

Regrettably, film sensitivity cannot be made to increase indefinitely by increasing emulsion-crystal size. There is a limit to how large a crystal can become and remain suspended in a gelatin solution. Crystals larger than about 4 microns are not of much use in

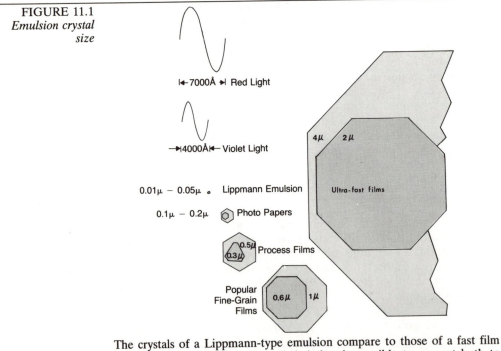

FIGURE 11.1
Emulsion crystal size

|←7000Å→| Red Light

→|4000Å|← Violet Light

0.01μ – 0.05μ ₀ Lippmann Emulsion

0.1μ – 0.2μ Photo Papers

0.5μ / 0.3μ Process Films

Popular Fine-Grain Films   0.6μ 1μ

4μ 2μ

Ultra-fast films

The crystals of a Lippmann-type emulsion compare to those of a fast film like pebbles to boulders. Indeed, it is barely possible to present both to scale on the same page. The crystals in Lippmann-type emulsions are smaller than a wave of light. They cannot be resolved by optical microscopes.

ordinary gelatin emulsions as they tend to sink during precipitation. There is also a limit to how thick and turbid the emulsion can be allowed to become in accommodating the larger crystals. Even without these complications, little would be gained in film speed from using crystals larger than about 2 microns. Light sensitivity stops increasing in crystals larger than this, and then for some reason declines. Also, if crystals as large as 2 to 4 microns are present in an emulsion, light from different parts of the subject may strike them and create multiple latent image centers. The development patterns that arise from crystals exposed in this way are somewhat unpredictable. Parts of the crystal may develop at different rates and other parts may not develop at all.

Emulsion sensitivity, or film speed, and resolving power are inversely related, each being affected oppositely by a change in the average size of emulsion crystals. But the inverse relation between them is not always strict. Film speed, for example, is as much dependent on chemical doping and sensitization as on crystal size. Two films may have identical speed ratings and quite different resolving power.

Emulsion crystal size ultimately determines the graininess of a film. Graininess is the mottled, mealy effect seen when photographs are enlarged. It degrades the smoothness and continuity of image tones and lowers the range of tonal gradation. It is minimized in films with small emulsion crystals and tends to increase as the average size of emulsion crystals increases. Graininess is not restricted to black-and-white photographs nor to any particular film type. It occurs to some extent in all silver-halide emulsions and is always manifest under sufficient magnification. The graininess of color films and of chromogenic black-and-white films is influenced by the properties of the color couplers used. Color couplers are discussed in Chapter 14.

Graininess

The size of the grainy particles seen in enlarged photographs is only partly accounted for by the size of the silver specks of individual silver-halide crystals. The crystals, as we have seen, are no larger than 2 to 4 microns across and in most emulsions are just a fraction of that. Even when a film is enlarged by 20 diameters, the silver speck created by the largest crystals will be smaller than can be discerned by the eye.

Graininess increases more significantly from the clustering of crystals that occurs during emulsion precipitation and ripening. Silver-bromide crystals have a natural affinity for one another. When free to move about in the emulsion, as they are during physical ripening, they tend to gravitate toward neighboring crystals and form clusters. This clustering, which worsens during the extended ripening period used in making coarse-grain, high-speed emulsions, is the significant source of graininess.

Crystal-Size Range

If all crystals in an emulsion were equally large, they all would be equally sensitive. All would produce latent image centers and become developable at the same exposure level; all exposed below that level would remain undevelopable. Such an emulsion can produce two density levels and no more. In areas where image brightness rises above the threshold exposure, all crystals become developable, reaching full density with full development. Elsewhere, where exposure falls below threshold, no crystals become developable and image density is at a minimum.

To produce images of wide tonal gradation, an emulsion must contain a variety of crystal sizes. Consider an emulsion that contains crystals of large, small, and intermediate size. In areas where the optical image is bright enough to create stable latent images in all crystals, including the smallest and least sensitive ones, maximum density will occur. In areas where image brightness falls just below maximum, the large and intermediate crystals will respond, but the smallest crystals will not. The negative will be dense in those areas, but density will be less than maximum. In areas where image brightness declines further, negative density will decline proportionally. In this way variation of tonality will occur with variation in image brightness.

As seen in Chapter 10, tonal gradation consumes part of the information capacity of a film; in designing an emulsion, resolving power and gradation can be increased concurrently only if information capacity is increased. One can see intuitively why this must be. To yield superior resolving power, emulsion crystals must be as small as possible. When they are, all will be of the same minimum size without exception. An emulsion composed of such crystals will have impressive resolving power, but will only reproduce high-contrast images without tonality. Improving the gradation of the emulsion means replacing some of the small, high-resolution crystals with larger ones. Since the use of the larger crystals increases the size of elemental areas, resolving power must decline as gradation improves.

Emulsion Thickness

Thin-Emulsion Films

It can be seen from Figure 11.2 that light scatter begins near the surface of an emulsion and increases progressively as light penetrates more deeply into the emulsion. Since light travels a shorter course through a thin emulsion, there is less opportunity for light to spread. Thin-emulsion films tend to have superior acutance and resolving power. Thin-emulsion films are sharper also because they contain fine crystals. Films that contain ultra-fine crystals, regardless of emulsion thickness, are nearly devoid of turbidity.

Thick-Emulsion Films

Although thick-emulsion films may have lower acutance and lower resolving power, they dominate the film market, partly because they tend to be faster and partly because they are more tolerant to overexposure than thin-emulsion films are.

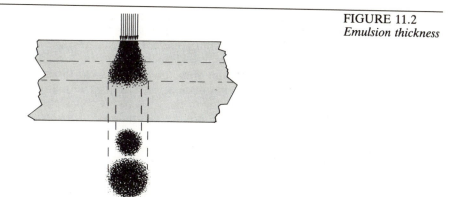

FIGURE 11.2
Emulsion thickness

The potential for image spreading is lessened in thin-emulsion films because there is less opportunity for light to spread than in thick-emulsion films. As a result, thin-emulsion films, in which the image records near the surface, produce sharper images.

To understand their tolerance for overexposure, think of a thick-emulsion film as having several thin layers or as being composed of several sheets of film stacked one on the other. When such a film is accurately exposed, the image records only in the top emulsion layer. If the film is overexposed moderately, highlights will record too densely in this top layer, showing little or no variation in tonality, but will record with variation in density in a subsequent layer. The negative, though dense, will retain tonal separation in diffuse highlights and it will be possible to print them for texture. Had you started with only the top layer or with a thin-emulsion film and overexposed by the same amount, you would have found it difficult or impossible to obtain comparable highlight gradation in the print.

The extent to which a film can be overexposed while retaining highlight separation or underexposed while retaining shadow detail is an indication of its *exposure latitude*. But you are forewarned that the exposure latitude given by a thick emulsion works only in the direction of overexposure. An *underexposed* film loses detail in shadow areas regardless of how thin or thick the emulsion layer is. If an image is too dim to record in the top layer, it certainly will not record in a subsequent layer. Also, exposure latitude is generally better in negative-type materials than in transparency or reversal materials. Reversal materials have virtually no tolerance for errors in exposure.

Of the factors that limit a negative's sharpness and resolution, the inherent properties of the film—its turbidity, speed, and granularity—ultimately decide the issue. Seasoned photographers do not seek to alter the characteristics of their favorite film when it fails to meet their needs, but from the variety of films on the market they choose one that best suits the assignment.

Considerations in Film Selection

The difficulty for the novice is in narrowing the choice, selecting one from among many competing films. The decision is complicated by the fact that a film's usefulness changes as conditions change. Today a slow, fine-grain film may be essential. Tomorrow a fast, coarse-grain film may save the day. So even though the film parameter of greatest importance in high-resolution films is resolving power, it is almost never the only factor considered in the selection. Other considerations are grain, gradation, and speed.

Grain Although the granular structure of the finest grained emulsions is unresolvable, even by the best optical systems, ultra-fine-grain films are much too slow and contrasty to be useful in ordinary photography. At the opposite extreme, modern high-speed emulsions have a coarse, granular structure that severely reduces resolving power and tonal continuity. To find a good, practical high-resolution film, one must search somewhere between the extremes of speed and resolution, between the extremes of the coarsest and the finest grain.

Gradation The gradation of a film is influenced by its developer as well as by its emulsion properties. Selecting a high-resolution, pictorial film is not a matter of choosing the film with the finest grain, but choosing the finest grained film capable of being developed for gradation. The limitation, which is discussed in Chapter 17, does not lie fully in emulsion technology but in finding better film-developer combinations. The potential for tonal gradation seems to peak near the grain-size range wherein emulsions first begin the transition from high contrast to gradation developable.

Emulsion Speed High-resolution films are usually slow, but this is not their strength. Emulsion speed is a useful property in any kind of photography provided it can be achieved without compromising image clarity. Fast films simplify the mechanics of high-resolution technique by allowing the use of faster shutter speeds to reduce motion blurring and smaller apertures to increase depth of field.

Black-and-White Films

Lippmann-Type Emulsions

If a film is to be selected for resolving power alone, if no other film parameter matters, Lippmann-type emulsions (named after Gabriel Lippmann, 1845–1921, a French physicist and Nobel prize winner) will be the films of choice. The crystals in Lippmann-type emulsions are extremely small, measuring just a fraction of a micron across. Being smaller than a wave of light and unresolvable under ordinary light microscopes, they went unseen until the invention of the electron microscope.

Kodak's high-resolution plates and films are Lippmann-type emulsions that resolve up to 2000 lines per millimeter or 50,000 lines per inch. One can see from Table 11.1 that the speed rating of Lippmann-type emulsions ranges from ISO 0.0025 to 0.01. The straightline

Film type	Lines per/mm	ISO speed	Remarks	TABLE 11.1
Lippmann-type emulsions	1000–5000	0.0025–0.01	Used in specialized scientific applications	*Film types*
Microfilms	200–2000	0.001–1	Used commercially in high-reduction copying	
Process films	100–500	0.1–64	Very fine grain; capable of high-contrast development or gradation development	
General-purpose films	30–100	25–4000	Resolving power declines rapidly as speed increases	

portion of the emulsion's characteristics curve is nearly vertical. They are extremely slow and contrasty. They are also characterized by the complete absence of turbidity, even in thick emulsion layers.

Lippmann-type emulsions that resolve up to 5000 lines per millimeter have been formulated, but they are so slow and insensitive that exposures may run several minutes in good lighting. During an exposure this long, the exceptional resolving power of the film can easily be lost to blurring from mechanical instabilities of the camera or in its support. This problem is sometimes solved by mounting the subject and the camera onto a common support of massive, rigid construction to ensure virtual immobility of the subject relative to the film.

Exploiting the incredible resolution potential of Lippmann-type emulsions can be an awesome task. Consequently, they have rather specialized applications, generally involving high-contrast subjects, wherein fine detail takes absolute priority over other considerations of cost and convenience, and wherein nothing else matters except resolving power. They are used, for example, to make masks for miniature microelectronic integrated circuits and to make reticles for optical instruments. They are also used to produce superfiche (microfiche at extreme reductions). In micrographic applications, emulsions like Kodak's high-resolution plates give legible reductions of 100 diameters or more, but they are of little value in pictorial photography because they cannot be developed for tonal gradation.

Microfilms

The high-resolution plates coated with Lippmann-type emulsions and the films used by the microfilm industry are similar. Both have extremely fine grain and are slow. Practical microfilms generally have greater speed but lower resolving power. Production microfilm strips and microfiche sheets are designed to provide convenient exposure times at reduction ratios of between 24 times to 48 times.

In high-contrast photography, line-copy work, and similar applications wherein tonal gradation is not needed, these films offer outstanding resolving power, but like Lippmann-type emulsions, they are too contrasty for pictorial photography.

Process Films

Process films are films designed to be used in graphic-arts reproduction, photoengraving, and in photomechanical offset printing. They are sometimes called copy films, high-contrast films, or lith films. Those sold in 35mm format go by trade names such as Kodak's Technical Pan Film 2415, DAF's Recording Film, and others. Process films are slow by modern standards, yet compared with Lippmann-type emulsions their speed is rather good. The essential demand imposed on films used in halftone applications is that they produce dots of high density and sharpness at convenient exposure speeds. In such applications process films are processed to yield high density and extreme contrast.

The characteristic of process films important to high-resolution photography is that they can also be processed to obtain the tonal gradation needed in pictorial photography. Even when process films are developed to produce maximum density, they may show signs of softness or gradation at edges. This softness proves that the film's inherent contrast is not infinite and that its emulsion contains a moderate range of crystal sizes.

When process films are developed in low-energy developers, they yield gradation negatives of very high resolving power. A film like Kodak's Technical Pan Film 2415 can produce prints of moderate legibility at enlargements of up to 25 diameters. When similar films are used skillfully, enlargements from 35mm negatives are barely distinguishable from enlargements made to the same size from high-speed 4 × 5-inch film. Their drawback is their fairly slow ISO 25 speed rating. (See Plate 5.)

General-Purpose Films

General-purpose films designed for pictorial photography—like Plus-X, Tri-X, and HP-4—are among the fastest and most popular films available. They are well suited to meet the needs of casual photographers who consume large amounts of photographic films. These films are designed with thick emulsion layers to yield extended exposure latitude and with high emulsion speed to allow for hand holding the camera. As shown earlier, however, the thick emulsion and coarse grain of general-purpose films trade off sharpness for exposure latitude and trade off image structure and gradation for ease of use. Although they are well publicized and promoted and are widely used, their strength is not in their imaging quality.

General-purpose films do produce acceptable images provided they are not enlarged by more than 8 or 10 diameters. In applications where small enlargement ratios will suffice and the negative will not be extensively cropped in printing, such films are useful. They are invaluable also in low-light surveillance photography where the use of slow, fine-grain films and long exposures would allow ground vibrations to degrade image clarity. In such situations the goal is often to attain marginal resolution, to record useful information about the subject even if definition is lost. In such applications fast films are indispensable.

Recent Advances in Film Technology

Chromogenic Films

Chromogenic films, introduced in 1980, are black-and-white films in which color-film technology has been exploited for improved image granularity. Film manufacturers have used the latest advances in color-coupler technology in these films to reduce graininess while maintaining high film speed. The image created by chromogenic films is composed of a dye instead of silver specks. Graininess is reduced because the dyes are packed together so closely that overlapping dye particles obscure the grainy structure. When chromogenic films are overexposed, even more of the color couplers are activated and graininess is further reduced. There is no evidence that this process improves sharpness or resolving power beyond that achieved by nonchromogenic films of comparable speed ratings.

T-Grain Films

T-grain films (the "T" means "tabular" for the flat structure of the crystals) have some remarkable properties. (See Plate 16.) Introduced in 1983, they contain wafer-shaped crystals that are thinner than ordinary crystals and that lie flat so that their largest projective surfaces face the imaging light. This orientation allows the crystals to absorb more light and increases the effective speed of the emulsion. Films like Kodak's T-MAX, introduced in 1987, feature tabular grains whose surfaces are textured or ruffled instead of smooth to further increase light absorption. Because the crystals are so thin, more of them can be put into a smaller amount of gelatin for improved information capacity. This, of course, yields greater detail and wider tonal gradation. Also, since the emulsion is made thinner, T-grain films generally give a sharper image.

Chapter 12

The Chemistry of Development
Forming Silver Images

Photographic chemistry, particularly the chemistry of development, is a subject filled with folklore. No topic in photography has given rise to as many contradictory claims. Many "miracle" developer formulas have been introduced over the years—to increase film speed, reduce grain, or improve sharpness—only to fizzle. Some have performed as promised. It is easy enough to find a developer good enough for high-resolution work, but to separate fact from folklore requires an informed approach.

The kinetics of development can alter an image in several ways that are of interest in high-resolution photography. The aspects of development examined in this chapter are the energy of development, the degree of development, and the special chemical properties of the developing agent.

Basic Photographic Processes

A photograph is formed by two essential chemical processes: development and fixation. During *development,* silver-halide crystals that have received enough exposure to have created latent image centers are chemically reduced to particles of pure silver. These particles of silver are the image specks. During *fixation,* the undeveloped crystals that were not converted to silver are chemically dissolved so that they can be removed from the emulsion. If left in the film, these sensitive crystals would, from the action of light alone, slowly reduce to silver and eventually blacken the entire image.

As image-forming activities, fixation and other postdevelopment processes are secondary. Their function is to desensitize the film, remove residual chemicals from the emulsion, and otherwise protect the image. The important aspects of image clarity—resolving power, acutance, and gradation—are established firmly when development ceases. The clarity of an image is not greatly affected by fixation and subsequent processing steps unless the emulsion is physically damaged. Although it is theoretically possible to alter the image, say by overfixing the film, it takes extreme action to do so. One might also reticulate the emulsion by allowing the temperature of the processing solutions to vary from one bath to the next. However, with reasonable caution during fixation and subsequent processing, image clarity will not be seriously affected beyond the development step. The focus in this chapter is therefore on image formation (the chemistry of development) rather than on fixation and postdevelopment processing.

Silver-Halide Reducing Agents

The key ingredient in a developer solution is the developing agent or the *reducing agent.* Do not confuse this kind of reducing agent with the kind used to bleach and reduce image density in overexposed prints. The kind of reducing agent used in development is one that removes the nonmetallic element from silver-halide molecules, giving up electrons in return. It is the reaction that follows the release of electrons that allows silver to be liberated from silver-halide crystals during development.

Many reducing agents that can react with silver halides to produce elemental silver are so powerful that they reduce exposed crystals and unexposed crystals indiscriminately. Such agents are useless as photographic developers. To have photographic value, a reducing agent must also be *selective*. It must show a disposition to act on exposed crystals without acting too vigorously on unexposed ones. Evidence of poor selectivity in a developer is a high fog level in the negative or a general increase in density (as in the unexposed borders of the film).

Developer Solutions

A *developer solution* is a mixture containing the reducing agent and other chemicals dissolved in water. One can conceivably prepare a developer using the reducing agent alone in the solution, but useful developer solutions are never mixed in this way. As seen below, other compounds are added for several reasons.

Antioxidants or Preservatives

A developing agent enters into a chemical reaction by oxidizing. Given the opportunity, it will react with oxygen in the air at the surface of the liquid or with the air dissolved in the solution. On its own a developing agent mixed in plain water will quickly lose its potency, exhausting itself prematurely before entering into a chemical reaction with an exposed crystal. Even when such a developer is freshly prepared before each use, a wasteful and time-consuming procedure, its action may be erratic from batch to batch. To prevent premature oxidation of the developer, preservatives are added to developer solutions.

Sodium Sulfite. Sodium sulfite is the major preservative used in developer solutions. It preserves the chemical activity and freshness of the developing agent, thereby prolonging the storage life of the solution. The tendency of developing agents to oxidize in water is effectively countered by replacing water with a 2 percent solution of anhydrous sodium sulfite. Alternatively, about 20 to 30 grams of sodium sulfite can be added to each liter of developer solution. Indeed, sodium sulfite is such a good antioxidant that it is used in developer solutions almost universally. Ascorbic acid and a few other compounds have preservative properties, but developers preserved with these compounds are rare.

Part of the reason sodium sulfite is so often used is that it does more than protect the developer from oxidation. You will see later how it affects a developer's induction speed and a film's exposure index. It plays a role in fine-grain development. It also neutralizes the brown developer stains produced by developer oxidation products. Finally, being mildly alkaline, it is an accelerator, as explained below, used to activate the developing agent.

Alkalis or Accelerators

Another reason developing agents are not used alone in developer solutions is that the common ones are inactive photographically in neutral or acid solutions. This is why acid stop baths and plain water baths are effective in stopping development. With isolated exceptions, developing agents must be used in an alkaline environment. Full developer activity is achieved by adding alkalis, such as those in Table 12.1, to the solution. As these compounds have the effect of speeding the action of the developer, they are called *accelerators.*

Of the many substances that provide an alkaline environment, the most useful are the buffered alkalis. Buffered alkalis release hydroxyl ions, whose quantity determines *alkalinity* or *pH value,* at a rate that keeps their concentration fairly constant over the active

TABLE 12.1 *Common photographic alkalis*	Alkali	pH value
	Sodium hydroxide, non buffering	13.0 (caustic)
	Sodium carbonate, buffering	10.7 (moderately alkaline)
	Sodium metaborate, buffering	10.2 (moderately alkaline)
	Borax, buffering	9.3 (weakly alkaline)
	Sodium sulfite, buffering	8.7 (very mildly alkaline)

life of the alkali. Nonbuffered alkalis simply lose their potency and decline in alkalinity with use. Buffered alkalis keep the activity of the developer solution at a more predictable level.

Fog inhibitors are added to developer solutions to improve the selectivity of the developing agent. Fog inhibitors are called *restrainers* because they restrain the action of the developer on those crystals that have received little or no exposure.

Fog Inhibitors or Restrainers

An ideal restrainer would inhibit the development of only those crystals that are unexposed or have been exposed below threshold, reducing image fogging without reducing image information. Actual restrainers are not this effective. They inhibit development in marginally exposed crystals that would develop if the restrainer were not present; they thereby force the use of a higher exposure index for the sake of achieving normal contrast in areas of low exposure. In effect, they reduce a film's effective emulsion speed.

Potassium bromide is an effective restrainer when Metol is the active developing agent. Organic restrainers like benzotriazole tend to reduce fogging more effectively than inorganic bromide restrainers with less reduction in emulsion speed.

Induction of Development

When exposed film is immersed into a developer solution, development does not begin at once. The initial buildup in density is negligible. The delay, which occurs both in papers and in films, can be observed in the way a paper print develops. During its first moments in the developer the paper is blank. After the image appears, however, density builds quickly. The delay in the rapid formation of silver is known as the *induction period*. The length of the induction period is influenced somewhat by the time needed for the solution to soak into the emulsion, but this seems to be a minor factor. When film is immersed into a developer the gelatin of the emulsion layer swells and soaks up the developer solution; the crystals are thus quickly surrounded by the active developing agent. A more significant time lapse occurs between the time the developer comes into contact with a crystal and the time silver is produced at the crystal in quantity.

Although it is not entirely clear why the rate of growth of the silver speck is so slow during the induction period, there are several possible explanations. Apparently the developing agent must be attached or adsorbed to the silver-halide crystal before it can reduce the crystal to silver. This adsorption seems to be facilitated by the presence of latent image centers on the surface of the crystal or just below the surface that provide points of attack at which the developer can react with the crystal. It may be that induction is slowed because latent image centers are inaccessible to the developer in the early stages of development. A latent image center may initially be buried in the interior of the crystal or it may be partially covered.

Perhaps surface latent image centers are protected by a barrier charge.

The delay in induction may also be affected by the small initial size of the latent image speck. The developer functions by transferring electrons to the crystal. Latent image centers seem to catalyze this transfer in proportion to their size. The larger the latent image center has become during exposure and the larger it grows during development, the more vigorous does the action of the reducing agent become. Since the electron affinity of a few atoms of silver may be less than that of a large mass, the catalytic properties of the small initial latent image center may simply be different during the early development period.

Whatever the cause, silver formation is practically dormant during the induction period and the increase in image density is small. Development commences after several seconds in heavily exposed crystals, after several minutes in lightly exposed crystals, and after several hours or days in unexposed crystals. The difference in the length of the induction period from one crystal to another seems to hinge most on the size, configuration, and accessibility of the latent image centers, whether they are on the surface or in the interior of the crystal.

The induction period is shortened by the presence of mild solvents, like sodium sulfite, in the developer solution. Sodium sulfite etches away at the surface of emulsion crystals enough to expose interior latent image centers to the developer.

The length of the induction period is also influenced by chemical properties of the developing agent. The time needed by a developing agent to develop a film to completion, when expressed as a multiple of the length of the agent's induction period, yields a number that is fairly constant for that agent. This number, known as the *Watkins factor*, is 5 for hydroquinone and 30 for Metol. In other words, the induction period of hydroquinone, a long-induction developer, is one-fifth its total development time. The induction period of the rapid-induction developer Metol is only one-thirtieth its total development time. When developer activity is adjusted so that the time needed to develop a film by a Metol developer equals that needed by a hydroquinone developer, silver will be produced by the Metol developer much earlier in the development cycle.

Energy of Development   Once the production of silver starts, it may proceed vigorously or at a controlled rate depending on the *energy of development*. The energy of development is affected by the reduction potential of the developing agent, by its concentration in the developer solution, by the temperature of the developer, by the method and extent of agitation during development, and more significantly, by the alkalinity of the developer solution as determined by the pH value of the alkali and its concentration in the solution.

Studies conducted on large silver-bromide crystals 15 to 17 microns in diameter show that development proceeds along different courses as the energy of development changes. During normal development, silver forms near the edge of a crystal and spreads around its surface. During more energetic development, the production of silver begins at several points on the crystal and spreads in several directions simultaneously. In some cases explosive development occurs. Silver is actually thrown from the crystal with such force that it propels the crystal through the emulsion.

Energetic development can contribute in a similar way to image-point spreading. It seems that the more energetic development is, the more likely it becomes that silver will extrude and spread beyond the original boundaries of the crystal. The retention of a film's resolving power and inherent image quality seems best ensured by regular, controlled, low-energy development.

Induction Speed Versus Development Speed

Induction speed and development speed are different phenomena, each being affected by different characteristics of the developer solution. In a rapid-induction, low-energy developer, silver appears early, but density increases more slowly than it would in a more active developer. Even in low-energy developers, however, once the production of silver at the exposed crystal begins, it proceeds swiftly to completion.

Some of the properties of a developer solution are accounted for by differences in the way induction speed and energy of development affect the speed of silver production. All developers, for example, act first on the heavily exposed crystals associated with brightly illuminated highlight areas. Long-induction developing agents are somewhat slow to act in highlight areas and are slower still in shadow areas. Once crystals in highlight areas start to develop, highlight density builds quickly and may become excessive before the production of silver in shadow areas proceeds very far. When highlight density is ideal, shadow density will be thin and weak.

Rapid-induction developers, on the other hand, act quickly on crystals in highlight areas and take proportionately less time to act on crystals exposed at lower brightness levels. Before density in highlight areas reaches an optimum level, shadow areas will also have begun to build density and will contain substantial detail.

One can therefore deduce that the induction speed of a developer affects the exposure index of films developed in it. Film speed is determined by the exposure needed to gain usable density in important shadow areas of a negative. Rapid-induction developers, by producing good shadow density, yield good emulsion speed. Long-induction developers give weak shadows that can be strengthened either by extending development or by increasing exposure. The first alternative, extending development, works poorly. It improves shadows but makes highlights far too dense. To get good shadow density as well as correct contrast and tonality with a long-induction developer, exposure must be increased. Highlight density is then moderated by curtailing development. The need to increase exposure, however, is equivalent to reducing the film's exposure index or using a slower film. The full emulsion speed of a film is favored by the use of rapid-induction developers.

Development and Resolving Power

Resolving power is affected by the degree of development, but during regular, low-energy development, it is affected in a small way. Resolving power declines most during a high-energy or extended development that increases image contrast above optimum.

In the typical development pattern of a film, resolving power increases during the first few minutes as the image builds density and contrast, reaches a maximum early in development, and then declines. Regrettably, peak resolving power occurs too early, before density reaches its optimum level. After resolving power peaks, it declines sharply, but its rate of decline quickly levels off. As development proceeds, resolving power declines at an ever slower rate. Figure 12.1 illustrates the typical pattern. Chemical spreading accounts for the gradual reduction in resolving power beyond the peak resolving power shown in Figure 12.1.

Developing Agents

The developing agents in widespread use today have good induction speed, form part of a superadditive combination, or are capable of reacting with certain chemicals to produce color dyes. Those that

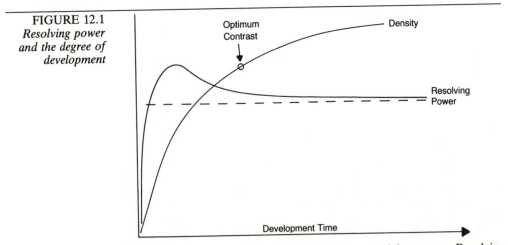

FIGURE 12.1
Resolving power and the degree of development

Development time has little practical effect on resolving power. Resolving power increases early in development as contrast builds in the image, but peak resolving power occurs before usable density is reached. Resolving power declines somewhat if the film develops beyond its optimum contrast, but the decline is small.

produce dyes are discussed in Chapter 13. The developing agents commonly used in black-and-white developers are: hydroquinone (p-dihydroxybenzene), Metol (n-methyl-p-amino-phenol), and Phenidone (1-phenyl-3-pyrazolidone).

Hydroquinone

Hydroquinone is a long-induction developer that builds high density in high-exposure regions but poor density in shadow regions. Despite this weakness, it is a widely used developing agent whose popularity originates from the regenerative and preservative effect it has when combined with the rapid-induction developing agents Metol and Phenidone, and from the synergistic properties it confers on the resulting developer.

Hydroquinone, as such, does not act as a primary developing agent in developer solutions (not even in solutions where it is used alone). Nearly all of the activity of a hydroquinone-based developer is caused by the presence of dianions: anions of hydroquinone with double negative charges. These dianions are not completely formed below a pH of 11.5 to 12; consequently, they are not active in typical developers accelerated by sodium carbonate and milder alkalis.

Hydroquinone is used as the sole reducing agent in applications where extreme image density or high contrast is needed. In these applications it is usually formulated as a high-energy, strongly alkaline developer, as in the D-8 formula. See Formula 12.1. Such a developer creates exceptionally sharp edges when used with high-contrast films, but it is not likely to bring out the best qualities in a turbid gradation film.

Metol

Metol is an excellent developing agent. It is a rapid-induction developer capable of producing very good shadow detail. It is highly selective and has moderate to low fogging characteristics. Used alone in a sulfite solution, as in a formula like D-23, Metol develops with very low energy, producing images of low contrast with tight grain. See Formula 12.2. In the D-23 formula it is a compensating developer useful in situations of built-in high contrast, as in scenes illuminated by street lighting or scenes containing bright light sources.

The addition of a weak bromide solution to a Metol developer lowers its tendency to fog the negative. Since fogging in Metol

Sodium sulfite	90	grams
Hydroquinone	45	grams
Sodium hydroxide	37.5	grams
Potassium bromide	30	grams
Water to make	1	liter

FORMULA 12.1
Caustic hydroquinone developer

Kodak's D-8 is a high-energy, long-induction hydroquinone developer that yields maximum emulsion density, high contrast, and excellent sharpness, but less than optimum resolution. It is activated by sodium hydroxide, a powerful, caustic, and dangerous alkali.

FORMULA 12.2 *Compensating developer*	Metol	7.5 grams
	Sodium sulfite	100 grams
	Water to make	1 liter

This formula is designated D-23 by Kodak. It is a soft-working compensating developer useful in high-contrast lighting situations, as in available-light photography, scenes illuminated by street lighting, and scenes containing light sources. A Metol-sulfite developer, it is one of the simplest practical formulas known.

developers is low to begin with, some Metol-based developers, like D-23, are formulated with no bromide at all.

Bromides, which restrain the activity of Metol more than they restrain Phenidone or hydroquinone, are responsible for the *adjacency effects* discussed in Chapter 17. It is this restraining action also that limits the useful activity of Metol-based developers. Since silver bromide is the silver halide most commonly used in print papers and films, bromide ions are usually released in large quantities during development as a by-product of the reduction of silver bromide. Developer solutions repeatedly reused, like those in commercial photo-finishing laboratories, eventually accumulate an excess of these bromide ions. Even when properly replenished, the activity of a Metol-based developer will eventually be curtailed by this bromide accumulation. There is thus a limit as to how often Metol-based developers may be reused.

Superadditive Combinations.   Metol is frequently used in combination with hydroquinone to form what are called MQ developers. The designation MQ simply indicates the presence of Metol and a quinone (hydroquinone) in the solution. Together, these developing agents constitute a superadditive combination that has synergistic properties, properties more useful than the sum of the properties of both agents used separately. In such a combination, development in highlights is greater than with hydroquinone alone and the induction period is shorter than with Metol. The superadditive combination offers good shadow speed as well as solid highlight density.

MQ developers depend on Metol rather than hydroquinone for developer activity. The importance of hydroquinone in the chemical reaction lies in its ability to regenerate Metol and restore Metol's activity. It chemically converts the oxidized by-products of Metol back into active Metol. As long as hydroquinone remains in the developer solution, the original quantity and activity of Metol does not diminish. On the other hand, the volume of hydroquinone declines as the developer is used.

Phenidone   Phenidone is a developing agent whose contrast is lower than that of Metol and whose induction speed is faster. It gives a somewhat greater exposure index than Metol gives and produces good density in shadow regions. It builds weak highlight density, however. Like Metol, indeed to an even greater extent, it is superadditive with hydroquinone. Phenidone is an active or potent developing agent capable of initiating development when highly diluted. Metol, by comparison, must be present in a developer solution at 10 to 20 times the concentration of Phenidone to generate equal activity. Phenidone is therefore an economical developing agent, although it costs two or three times more than an equal weight of Metol.

Phenidone is used in many of the same ways Metol is, but because it is still being exploited commercially, few Phenidone formulas have been published. Excellent Phenidone-hydroquinone developers, known as PQ developers, can nevertheless be formulated by substituting Phenidone for Metol in MQ formulas.

Usually two modifications are made when an MQ formula is converted to PQ. First, because of Phenidone's activity, its weight can be reduced to about one-tenth that of Metol to retain a comparable developing time. Second, PQ formulas contain organic fog inhibitors like benzotriazole either as a replacement for, or in addition to, bromide inhibitors. Phenidone has a stronger tendency than Metol to fog the image, and bromides are less effective in restraining its fogging.

Phenidone has unique properties—when it is substituted for Metol in MQ formulas it may not always produce a comparable developer. In a Metol-sulfite developer like D-23 where Metol is the sole developing agent, Phenidone does not reproduce D-23's properties. A Phenidone-sulfite developer used in combination with general-purpose films cannot build usable highlight density even with the contrasty scenes that D-23 handles so well. Used without the benefit of hydroquinone for superadditivity, Phenidone produces so little highlight density that such a developer was long thought to be useless.

Not until 1967 was an application found for the soft-working Phenidone-sulfite developer. Marilyn Levy of the United States Army Electronics Command demonstrated a mixture, which she designated POTA, to be useful in wide-latitude photography or in conjunction with scenes of uncommonly high contrast. For example, POTA developers can hold detail in the fireball of an explosion while retaining detail in the background. POTA developers and developers derived from the POTA formula have since been used to obtain gradation development in very slow, ultra-fine-grain process films. Some of these are discussed in Chapter 17.

Chapter 13

Color Processes
Forming Dye Images

Color films are similar to black-and-white films in that both contain silver-halide emulsions, respond to light in the same way, and are capable of producing silver images. They are different in two important ways: color films have three emulsion layers instead of one, and the final image in a color film consists of organic dyes instead of elemental silver. Because of these differences, the resolving power of color films may be lower than that of black-and-white films of comparable speed. To understand why, one must understand something about color theory and about the chemistry of color development.

Color photography as it is now practiced confirms a hypothesis advanced in 1802 by English physicist Thomas Young (1773–1829), who thought it unnecessary to use every conceivable spectral color to reproduce a color image. According to Young's theory of color perception, the visual sensation associated with a full spectrum of color can be duplicated using as few as three fundamental colors. Indeed, color images can be synthesized using what are now known as the primary colors (red, green, and blue) or their complements (cyan, magenta, and yellow).

Using an appropriate set of color filters, one can record a full-color image on three sheets of black-and-white film. Color-separation negatives of archival permanence are made in this way. Making separation prints is a bit complex, however, and requires three separate exposures, perfectly aligned onto the same sheet of paper, one from each negative exposed through its original filter. A slight error in aligning these three images can degrade image sharpness considerably.

In modern color photography the three images are recorded on an integral tripack—a film that contains three separate emulsion layers on a single base. Color films have nonemulsion layers as well: one is an overcoat layer that protects the emulsion layers from abrasion, and another is a filter layer that helps preserve color purity. An anti-halation layer may also be incorporated to absorb excess light.

The three emulsion layers are designed so that the silver-halide crystals of each are spectrally sensitized to respond to light of one primary color. The spectral sensitivity of an emulsion is altered when special sensitizing dyes are adsorbed to the emulsion crystals. Nonsensitized emulsion crystals, crystals in their natural state, react to blue and ultraviolet light but are insensitive to light of longer wavelengths.

The top emulsion layer of a typical color-negative film reacts to blue light, the middle emulsion layer responds to green light, and the bottom emulsion layer responds to red light. The two bottom layers, although spectrally sensitized to green and red light, retain their inherent sensitivity to blue light as well. A yellow filter layer is therefore needed below the top layer to absorb blue light, allowing only red and green to pass to the layers below.

Color Couplers

The image in a color photograph consists of dyes, but outside of dye-destruction processes like Cibachrome™, the dyes are not built into the film from the start. Each emulsion layer contains a coupler that has the chemical composition needed to produce a dye image of a color complementary to the color of the exposing light the layer has been sensitized to record. The top layer contains a yellow coupler which yields a yellow dye in exposed areas upon development. The middle layer contains a magenta coupler. The bottom layer contains a cyan coupler. Color couplers are, in effect, half the dye molecule. The integral tripack owes its existence to these couplers— various phenols, naphthols, and pyrazololones—that react with the by-products of certain developing agents to create dyes of the correct color.

It is the chemical composition of the coupler that determines the final dye color; this is why each emulsion layer must contain a unique coupler to create the yellow, magenta, or cyan dye needed in that layer. The chemical composition of the developing agent is just as important. A color developer, in addition to producing silver, must oxidize to create specific by-products that have the chemical composition needed to complete the process of dye formation. These by-products are in effect the second half of the dye molecule. Not every developing agent produces them. Paraphenylene diamine and paraminophenol, two notable exceptions, are highly selective, rapid-induction developers capable of color development. Metol and Phenidone are essentially black-and-white developers; they produce silver quite efficiently, but their reduction by-products do not react with color couplers to produce dye.

Dye Formation

Dye formation breeds imaging problems that are unique to color photography. One such problem arises from the size of the couplers. In early attempts to create an integral tripack, the coupler molecules were small compared to the molecular structure of gelatin in the emulsion layer. As a result, couplers were mobile and tended to diffuse freely, occasionally wandering into adjacent emulsion layers. Since it is the chemical composition of the coupler that determines the color of the image dye, the presence of the wrong coupler in an emulsion layer contaminates color purity by creating the wrong dye in that layer. One early problem posed by the integral-tripack approach to color photography, one that had to be solved before color purity and image accuracy could be achieved, was containing the dye-forming chemicals in the layer in which they were originally placed.

The problem has been solved in different ways by various film manufacturers. Heavy molecules of oil or resin are adsorbed to the couplers of some films, literally anchoring the couplers to the emulsion layer. In other films, couplers are selected according to their molecular size and mass. Couplers with long carbon-chain structures stay neatly in place because they are too massive to move. One film manufacturer uses an altogether different strategy: the couplers are omitted from the film entirely and are introduced into the emulsion during processing, one layer at a time.

Dye-Cloud Formation

The initial chemical processes that characterize black-and-white development characterize color development. In both cases, image formation begins with the action of a developing agent that produces metallic silver at the site of exposed crystals. In black-and-white processing the silver image is final. In color processing the silver image is temporary and is eventually dissolved from the emulsion.

The important chemical activity in color development occurs simultaneously with the formation of the black-and-white silver image. This activity includes the reaction mentioned earlier between couplers and the oxidized by-products of the developing agent. Color dyes are formed, as stated, when color couplers combine with exhausted developer. The fact that this exhausted developer is produced only at sites where crystals are undergoing development means that it will react with couplers only in areas associated with the image. Consequently, dye will be formed only in areas where exposure has occurred.

The way a color developer combines with color couplers to create dye has an important influence on the resolving power of color films. As the developer reduces exposed emulsion crystals to silver, molecules of oxidized developer diffuse outward from the development site until they encounter and chemically combine with an available coupler molecule embedded in the emulsion. Because of the outward diffusion of the oxidized developer, this chemical reaction occurs at various distances from the site of development. The aggregation of dye particles created in this process resembles a cloud that surrounds the developed crystal. From the way it is formed, it is clear that this dye cloud will always be larger than the original crystal. Dye-cloud formation and the chemistry of color development thus contribute in an important way to the spread function of color films.

Much research has been devoted to reducing the size of the dye cloud and to finding active color couplers. Active couplers combine quickly with the oxidized developer before it diffuses too far from the crystal site. Dye specks are therefore formed closer to the original crystal, reducing the spread function and improving image quality.

In color-negative materials the density of the dye produced at the sites of exposed crystals increases with exposure. Highlight areas gain in density while shadow areas remain clear. A reverse-toned, negative image is thereby created in one development step. The processing of color negatives is thus rather straightforward.

Color Negatives Versus Color Transparencies

Transparency materials require an additional development step to produce a positive color image. Color-transparency film is first processed in an ordinary black-and-white developer, one that does not react with the color couplers in the emulsion. This is the *first developer*. Several black-and-white MQ or PQ developers may be used in this step. They operate here as they do in black-and-white processing, creating negative silver images, a separate image formed for each of the three emulsion layers. Once formed, these silver images serve no further purpose and can be bleached out of the film. Since they do not interfere with the remaining processing steps, they are just as well removed at the conclusion of color development. The color couplers remain unaltered at this point.

As a consequence of the first developer, positive ghost images taking the form of unexposed and undeveloped silver-halide crystals are left in each of the three emulsion layers. These positive images correspond to areas that were not converted to silver by the original exposure and first development.

The three positive images are now processed in a fogging, color-forming developer to create the final dye images. During this step, called *color development,* several reactions occur simultaneously: silver-halide crystals are made developable by the fogging agent; the fogged crystals are reduced by the color developer, generating more silver and releasing oxidation by-products; and dye is created to form the three color images as the developer by-products react with color couplers at development sites.

When all is done, silver created in both the first developer and the color developer extends to all parts of the film. This silver is bleached out so that only the dye images remain.

Nonsubstantive Films

The problem of restricting the color coupler to its original layer is solved in one kind of film by leaving the couplers out of the emulsion layer altogether and introducing them during processing. Such films are called *nonsubstantive*. The film is processed one layer at a time and the coupler needed in that layer is introduced into the developer solution during three separate color-development steps. Other films with incorporated couplers are called *substantive*.

The essential requirement in color development is that the appropriate coupler be available at the site of a developing crystal at the time of development. The way couplers are made available is important and can make a difference in a film's imaging properties.

There are some important advantages to the nonsubstantive approach. First, there is no need in nonsubstantive films to make the coupler molecules heavy and immobile; soluble couplers with a small molecular structure are used instead. The dye particles formed by soluble couplers are substantially smaller than those formed by incorporated couplers. They thereby reduce the size of the dye cloud and reduce the emulsion-spread function.

Second, although the soluble color couplers react with the oxidized developer in much the same way as incorporated couplers do, being mobile they diffuse toward the crystal and take part in the reaction closer to the site of the developing crystal. The inward diffusion of the soluble couplers compensates partly for the outward diffusion of the oxidized developer. Again dye clouds become smaller and denser.

Finally, the fact that massive color couplers are not incorporated into the emulsion means that the emulsion layer can be made thinner. Therefore, nonsubstantive films have lower turbidity and improved image sharpness.

Some of the nonimaging properties of nonsubstantive films are superior as well. Because the emulsion has no perishable organic couplers, nonsubstantive films, which are essentially multilayered black-and-white films, have better keeping properties and longer shelf lives. The incorporated couplers of substantive films may deteriorate during storage. The couplers of nonsubstantive films are added freshly upon processing.

Also because the need for massive coupler molecules has been eliminated in this process, film manufacturers can choose from a larger variety of couplers, selecting one for reasons other than size. Indeed, couplers used in nonsubstantive films are selected according to the stability of the dyes they form. As a result, nonsubstantive films have fade-resistant dyes with stable image characteristics.

Processing Nonsubstantive Films

Despite the superior advantages of the nonsubstantive process, the Kodachrome™ family of films from Eastman Kodak are the only nonsubstantive still films presently marketed. See Figure 13.1. The reason is that these films are difficult to process. The complex

K-14 process used with Kodachrome is one of the few color processes not used in home darkrooms. In fact, K-14 processing is beyond the capabilities of many commercial processing laboratories.

In the K-14 process the exposed film is first developed in a black-and-white developer to produce negative silver images and positive silver-halide images in each emulsion layer. Most of the remaining processing steps affect the positive silver-halide images. The film is exposed to red light from the base side, which activates the red-sensitive emulsion in the bottom layer without affecting other layers. It is then developed in a color developer to which a soluble cyan coupler has been added. This produces a positive cyan image in the bottom layer. The film is next exposed to blue light from above, activating the top blue-sensitive layer. A yellow-filter layer below the top layer prevents the green-sensitive middle layer from being affected by the blue-light exposure. The film is developed again, this time with a yellow coupler dissolved in the developer to produce a yellow image in the top layer. After the top and bottom layers have been exposed and developed, only the middle layer has light-sensitive silver-halide crystals in it. This layer is developed with a fogging developer containing a magenta coupler to produce the magenta image. Finally, the silver is bleached out, the positive images are stabilized, and the film is washed and dried.

Interspersed among these steps are several wash and stabilization baths. It is this complexity in the processing of nonsubstantive films that discourages manufacturers from making and marketing other films of this kind. Substantive films, being easier and less costly to develop, are widely available, but they sacrifice imaging quality for their economy and simplicity.

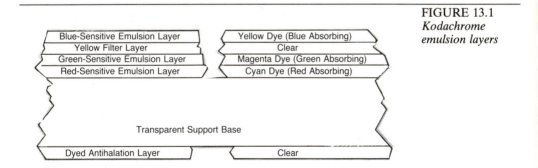

FIGURE 13.1
Kodachrome emulsion layers

Kodachrome, a film produced by the Eastman Kodak Company, is a non-substantive film whose thin emulsion layers contain no color couplers. Although nonsubstantive films have many superior image qualities, developing these films is an involved process.

The Resolving Power of Color Films

The principal limits on the resolving power of color films arise from the imperfections of their color couplers and dyes, from the nature of the dye clouds formed during development, and from the thickness of their emulsion layers. The fact that color materials contain three sensitive emulsions and several filter layers means the emulsion layer must be thick. Emulsion thickness is even greater in films having massive incorporated color couplers. As a result, the turbidity of color materials is typically greater than that of thin-emulsion black-and-white films.

A color image does not have to be as good as a black-and-white image to seem as sharp and clear. Color itself adds contrast and improves the perception of sharpness—to an even greater extent than tonal contrast does. Edge sharpness will seem greater in color photographs than in black-and-white photographs with the same objective sharpness.

The layering sequence of a color image also affects visual clarity. The magenta layer contributes most to the visual sensation of sharpness. Its position in the emulsion determines how much it will be degraded by light scatter. If magenta is coated as the top emulsion layer, light scatter in this important layer will be minimized and the sensation of sharpness will be improved. The yellow layer, which contributes least to the sensation of sharpness, is best coated at the bottom of the emulsion pack where turbidity is greatest. The layering pattern that yields greatest sharpness is thus magenta-cyan-yellow. Yet as you have seen, this is not the layering pattern used in typical color emulsions. Other practical considerations prevail in determining the layering sequence. In some paper emulsions, for example, the cyan layer is coated on top because it is the most fade resistant of the color dyes. When the cyan layer is coated above the less stable dyes, color prints tend to have a longer life.

Choosing a Color Film

Whenever conditions allow use of a slow film, the Kodachrome films are a good choice for color images. Because they benefit from the many advantages of the nonsubstantive process, they are among the best color films available.

In deciding on a color film, keep in mind that high-resolution, fine-grain color films have limitations similar to those of high-resolution black-and-white films: low emulsion speed and narrow exposure latitude. To exploit the full capabilities of the ISO 25 and 64-speed Kodachrome films, they must be used with a tripod-supported camera and must be accurately exposed. Also, color films have a grainy structure that tends to worsen as film speed increases. Color-reversal or transparency films like Kodachrome, unlike negative materials, have little latitude for overexposure. Most of the latitude of color-reversal materials lies on the side of underexposure and they have precious little of that.

Certain high-resolution assignments can be completed successfully without color film. On critical assignments the superior resolving power of black-and-white films will predispose the cautious photographer to select a fine-grain black-and-white film; a color film should be selected when recording the colors themselves is essential to the assignment.

Chapter 14

Photographic Lenses
The Magic of Fine Optics

The technology of lens design and manufacture has advanced to the point where it is often the film that limits the resolving power of a photographic system. A well-made lens certainly performs better, on the whole, than a fast, coarse-grained film. When one switches to an ultra-fine-grain, high-resolution film, however, the situation can easily be reversed. In high-resolution photography, it pays to select the finest lens available, and it pays to know how to select it.

Many modern lenses are more than adequate for general, non-critical use, but lenses of good construction, made by reputable manufacturers, vary widely in their imaging quality. Selecting one for high-resolution work requires a cautious approach. You must realize that resolving power is rarely the only parameter or even the dominant parameter in lens-design decisions. Also, the lens affects imaging quality in other ways—by its susceptibility to flare and by depth of field.

Lens-Design Limitations

Several factors prevent general-purpose lenses from being as good as photographers want them to be.

Optical Constraints

A lens can perform optimally at only one working distance. This is a constraint entirely different from the lens' limited depth of field (discussed later in this chapter). Based on the intended use of the lens, the designer must set specific lens-to-subject and lens-to-film distances, called the *optimum conjugate distances,* at which lens performance will be best. Most general-purpose camera lenses are optimized to perform best when focused at infinity. Their optimum performance is therefore never achieved when they are used to photograph subjects at finite distances.

The performance of general-purpose lenses is also reduced by the **Aberrations** presence of residual aberrations. Aberrations are the result of inevitable errors in lens design and construction. See Figure 14.1.

The aberrations of most concern here are those that increase the size of the spread function. Of the seven optical aberrations discussed below, all but one increase the spread function directly. Spherical aberration, coma, and astigmatism alter the distribution of light in ways that enlarge or distort image points. Chromatic and transverse chromatic aberrations cause color fringes that smear image points. Curvature of field and distortion alter the position of image points but do not enlarge them; however, curvature of field moves part of the image outside the plane of focus and in this way introduces image spreading. Distortion changes the position of image points, not their size or clarity.

Spherical aberration occurs because the focal length of a lens with spherical surfaces varies from the center of the lens out to the periphery. Axial and peripheral rays are thus brought into focus at different distances from the lens. See Figure 14.2. The degradation brought about by spherical aberration is lessened at small apertures because rays enter the lens close together proximally near the center; the difference in their focal points is thereby minimized.

Spherical aberration, even though it degrades image points, is exploited commercially. It is deliberately designed into certain soft-focus portrait lenses. A point affected by spherical aberration features a bright central spot surrounded by a less-intense halo (see Figure 4.1). The halo diffuses the image, giving a kind of softness

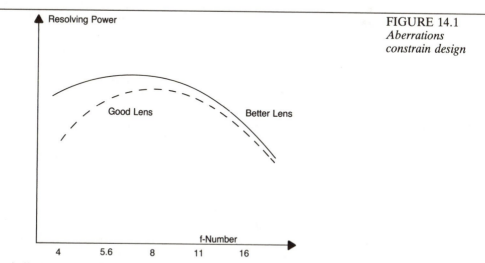

FIGURE 14.1
*Aberrations
constrain design*

The challenge to lens designers is to improve optical performance at large apertures. Performance at small apertures is constrained by diffraction, which cannot be mitigated during the design of the lens. Reducing aberrations is the important task of the optical engineer.

FIGURE 14.2
*Spherical
aberration*

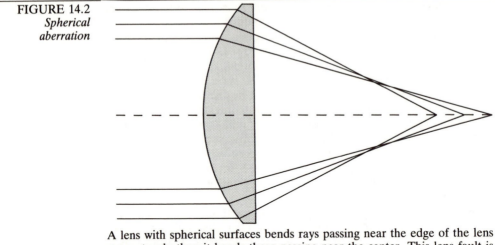

A lens with spherical surfaces bends rays passing near the edge of the lens more steeply than it bends those passing near the center. This lens fault is called spherical aberration.

that is pleasantly attractive and flattering in portraiture. At the same time, the small central spot gives fairly good image resolution. The effect is that of having a detailed image overlaid with one of soft, ethereal quality. Some photographers think that spherical aberration gives the best soft-focus images of all.

Astigmatism is an off-axis aberration that causes points to reproduce more like lines or elongated ovals than like circular disks. It degrades an image more at the edge of the field of view than at the center and can therefore be reduced by stopping down the lens. See Figure 14.3.

An astigmatic lens creates two separate focal planes, each at a different distance from the lens. The focal lines in one plane are oriented radially to the axis of the lens (radial lines, if extended, would cross the axis of the lens like spokes on a wheel). The focal lines in the other plane are oriented transversely to the axis, perpendicular to the first (when transverse focal lines are connected, they form a circle around the axis of the lens like the rim of a wheel).

Image lines are rendered most sharply by an astigmatic lens when they are precisely radial or precisely transverse to the lens axis, and then only when like focal lines are favored during focusing. It is impossible for lines of both types to be rendered sharply at the same time. Image lines that are not aligned radially or transversely with the lens axis cannot be focused sharply at all by an astigmatic lens.

Coma is an off-axis aberration that creates distorted, skewed image points. The focal spot of a point affected by coma resembles the oblique cross section of a point affected by spherical aberration (see Figure 4.1). Because this off-axis aberration worsens at points away from the axis of the lens, it is minimized at small aperture settings.

Both coma and astigmatism enlarge as well as distort image points. Both produce a disturbing, smeared image spot with none of the redeeming qualities of the image spot produced by spherical aberration. Coma and astigmatism are both especially troublesome at large aperture settings. They are, in fact, the lens faults that most seriously limit the resolving power of a lens used at or near its maximum aperture. They also become rapidly more severe as the field of view of a lens increases, making them difficult to eliminate from the design of wide-angle lenses. They make it extremely difficult to design wide-angle lenses that have large apertures—hence the scarcity and expense of fast, wide-angle lenses.

Chromatic aberration occurs in a simple lens because the refractive index of glass, or its bending power, varies with the wavelength, or the color of light. See Figure 14.4. Such lenses cannot bring light of different colors into focus at the same focal distance. When an image formed by one wavelength is in focus, the images formed at other wavelengths are blurred. Enlarged, out-of-focus spectral components account for the visible color fringe associated with this aberration. The damage done by chromatic aberration can be reduced somewhat by using small apertures to increase depth of focus, but chromatic aberration itself is unaffected by the aperture setting and is best eliminated during the design of the lens.

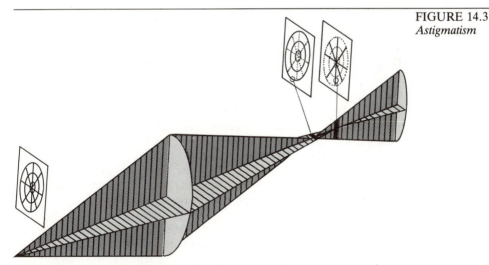

FIGURE 14.3
Astigmatism

Astigmatism is an off-axis aberration that creates in two separate planes focal spots resembling lines more than disks. Lines and edges are sharply rendered by an astigmatic lens only when they are oriented in the same direction as one of these focal lines and only when that focal line is favored during focusing. An astigmatic lens will render transverse lines and radial lines with equal clarity, but not at once.

FIGURE 14.4
Dispersion

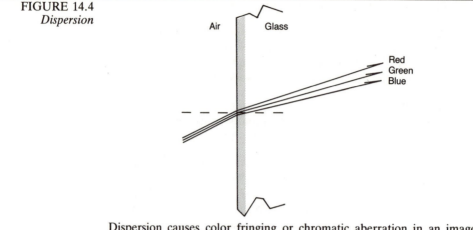

Dispersion causes color fringing or chromatic aberration in an image. It occurs in a simple lens because the bending power of such a lens varies with the wavelength of light passing through it.

Transverse chromatic aberration, also called lateral color, occurs when lens magnification varies with wavelength. This aberration can occur even after chromatic aberration has been corrected. Transverse chromatic aberration causes differences in the size of images of different colors. It splits a white-light image into several overlaid images, each of a different color and size. It thereby creates a rainbow fringe similar to that created by chromatic aberration. Changing the aperture setting can bring absolutely no relief from the effects of lateral color. It can only be corrected during the design of the lens.

The chromatic aberrations are the reason why telephoto lenses are more difficult to correct than lenses of shorter focal length. Because they magnify the chromatic aberrations and since small apertures do not eliminate these aberrations, long focal-length lenses must be well designed if they are to equal the resolving power of moderate focal-length lenses.

Curvature of field exists when the sharpest points in an image fall on a curved hemispheric surface instead of on a flat plane. It becomes impossible to focus the image sharply from corner to corner on film. If the center of the image is in focus, the perimeter will be blurred and vice versa. Curvature itself is not affected by the lens aperture. Because aperture affects depth of focus, however, it determines how badly curvature of field will degrade the sharpness of focus.

Distortion changes the relative position of image points. In the presence of this aberration, rectangular or cubical objects have curved instead of straight sides. When the sides curve so that their midpoints are closer together, the effect is called *pincushion distortion.* When the sides bend away from the center, the effect is called *barrel distortion.* Distortion is unaffected by the aperture setting and has no material effect on resolving power.

Production problems can prevent lenses from being as good as photographers want them to be. A simple, single-element lens or meniscus of spherical design (a magnifying glass, for example) suffers from all of the aberrations described above and makes a poor camera lens. Practical lens designs reduce these aberrations by combining lens elements, often in complex patterns, using different surface curvature and several kinds of glass.

Such lenses must be ground and assembled with meticulous care. Yet to control production costs, manufacturers may be willing to compromise construction quality by accepting liberal tolerances for assembly errors. It is not uncommon to find significant decentering in one of every three or four lenses coming off a lens assembly line, even in reputable plants. Yet if the center of any lens element is displaced from the optical axis of the lens by more than a few microns, its imaging quality will be seriously degraded.

Manufacturing Errors

Lens makers must also be sensitive to consumer demands. For example, a typical novice photographer wants to use the same lens to photograph subjects several inches away or several miles away. He likes lenses that have large maximum apertures and exotic focal lengths. He avoids lenses that are large, bulky, or heavy. He also expects lenses to be available at reasonable cost. These expectations greatly complicate the process of designing a lens and are another reason why resolving power may not be the critical parameter in the design of a general-purpose lens. Manufacturers cannot maximize resolving power, satisfy consumer demands, and remain competitive with one design. Demands for faster, lighter, and lower-priced lenses serve as design constraints for the optical engineer whose goal becomes not creating the sharpest lens, but creating the best compromise in performance that satisfies all constraints. Under such pressures, resolving power will matter only when it can be improved within design parameters.

Consumer Demands

The assumptions used in the design of high-resolution lenses differ from those used in the design of general-purpose lenses. General-purpose lenses must perform adequately in a variety of situations, at different subject distances, and at different apertures. High-resolution lenses, on the other hand, are specialized, intended to be used under known conditions over a limited range of subject distances. Such lenses—whether enlarger lenses, process lenses, or close-up lenses—are intended for such limited applications that they present the designer with few design constraints. The aberrations of specialized lenses can be corrected almost completely according to the specific conditions in which they will be used. Process lenses, for example, are used with flat copy, rather than with three-dimensional subjects, and are used at lens-to-subject distances of only a few feet. When a process lens is optimized for just these conditions, it can be designed for superb performance, especially if the eventual cost and size of the lens are not design constraints.

High-Resolution Lens Designs

Lens Flare

Lens flare, although it does not increase point spreading, degrades image clarity by reducing contrast, which reduces the visibility of fine features and lowers resolving power. Flare consists of unwanted light that manages to enter the camera to strike and expose the film. It cannot be entirely eliminated and it reduces image contrast disproportionately to its magnitude. Consider what happens when a unit of flare reaches the film plane. If illumination in the image initially ranges from 1 unit in the dimmest area to 99 units in the brightest area, the contrast ratio will be 99:1 if flare does not occur. Adding one unit of flare increases illumination to 2 units in dim areas and 100 units in bright areas. The contrast ratio drops to 50:1. One unit of flare has cut contrast in half.

Flare invariably weakens the strength and clarity of detail in shaded areas and can eradicate marginally resolved textural details elsewhere in the image. This loss of image detail is not just theoretical but real. Flare destroys image detail as predictably as bleaching.

Much of the stray light that comprises flare enters a camera through the lens. It can come from light sources within as well as outside the lens' field of view.

Flare from External Light Sources

Light that strikes the front element of a lens from outside the field of view can enter the camera as flare as it reflects off the lens barrel and internal surfaces of the camera body. This source of flare can be reduced by lens hoods.

Solar Flare

Bright sunlight creates a strong, obvious flare pattern when the sun is within or is just outside the field of view of the lens. This pattern consists of several repeating ghost images of the diaphragm created when sunlight is partly reflected at air-glass interfaces. (See Plate 6.) The pattern may be accompanied by light streams, streaks, smears, and color fringes. The intense brightness of sunlight makes its flare pattern clearly visible over the dimmer image light. Yet solar flare is merely a special case of the general phenomenon of flare. In general, each image point creates its own flare pattern to reduce the contrast and detail of an image in a manner that differs from solar flare only in intensity.

Flare from Image Light

The light that reaches a lens may be refracted at the surface, reflected, or absorbed. The part refracted passes through as it should, as image light. The part absorbed is lost and has little effect on the image. The part reflected can be damaging. If light is reflected away from the lens it will simply be lost—except for a slight dimming of the image, it will have no important effect. Image light reflected by internal lens surfaces, however, can enter the camera as nonimage flare light.

When light passes through an uncoated lens, the fractional amount reflected at each lens surface can be determined from the relationship:

$$\left(\frac{n_1 - n_0}{n_1 + n_0}\right)^2$$

where n_1 is the refractive index of the first medium and n_0 is the refractive index of the second. Using 1.5 in the formula as a typical value for the refractive index of glass and 1.0 as the refractive index of air, it can be seen that as much as 4 percent of the light reaching each air-glass surface will be lost by reflection. In a three-element lens this loss will occur at 6 air-glass surfaces. Total transmission by image light will then be:

$$(1 - 0.04)^6 = 0.78$$

Twenty-two percent of the light arriving at the lens will be scattered by reflection, much of it entering the camera as flare. The best remedy for this kind of flare is a coated lens. Modern lenses are coated with antireflection layers that increase lens transmission to 96 or 98 percent or more.

Flare is a general condition that affects all parts of the image, but it may not be distributed evenly over the image field. Depending on its source, it may be brighter near the edge of the film than at the center. It is better for this variation to occur in this way than for flare to concentrate at the center of the image. At the edge, flare partly counterbalances light falloff in the lens, which reduces image brightness from the center of the image out to the edge of the image field.

Focusing Errors

Several problems make critical focusing difficult. First, to get peak sharpness, the lens must be used at its optimum aperture, which has limited depth of focus. For example, when one focuses a 50mm lens on a subject 10 meters away, an error of 0.1mm at the film plane will shift the plane of sharpness at the subject by nearly 3 meters. Small errors in focusing can greatly enlarge the spread function. Second, it is difficult for the unaided eye to gauge whether focusing is critically sharp. The viewfinder image of a 35mm camera is so small that, even with moderate magnification, the eye cannot resolve the detail captured by a high-resolution system. The image could be badly out of focus, but one would be unable to tell.

Also, experimentalists may encounter a couple of exotic focusing problems: traveling focus and chemical focus. These problems are almost unheard of by photographers using modern films and well-corrected lenses and they are little more than of academic and historic interest today.

Traveling Focus

Traveling focus can exist in a lens having spherical aberration. Certain soft-focus lenses, for example, owe their softness to uncorrected spherical aberration left purposely in the design. In these lenses the aperture, which controls the amount of spherical aberration present, is used to vary image softness. But as the aperture of such a lens changes, the point of minimum confusion, the best focus, shifts, as illustrated in Figure 14.5. The standard procedure of focusing the lens at maximum aperture and stopping it down for the exposure will worsen image sharpness. Because of traveling focus, lenses with spherical aberration should be focused at the taking aperture.

Chemical Focus

Chemical focus can occur if a non–color-corrected lens is used with a non–color-sensitized film or paper. See Figure 14.6. A lens with longitudinal chromatic aberration brings each spectral color in an image to focus on a different focal plane, creating multiple images at different distances from the lens, a separate image for each color in the imaging light. The eye, being most sensitive to green and yellow light, focuses best on the image formed by yellow-green light. Untreated silver-halide crystals, however, respond only to blue and ultraviolet light. If the image is to be rendered sharply on film, the blue and ultraviolet wavelengths are the ones that must be in focus at the film plane. When the green and yellow wavelengths are focused visually, the blue and ultraviolet wavelengths will be blurred and the film image will not be as sharply rendered as supposed. Visual focus and chemical focus are at variance in such a system. To render the image sharply on film, the point of focus determined visually must be adjusted by focusing a predetermined amount beyond the point of visual focus.

FIGURE 14.5
Traveling focus

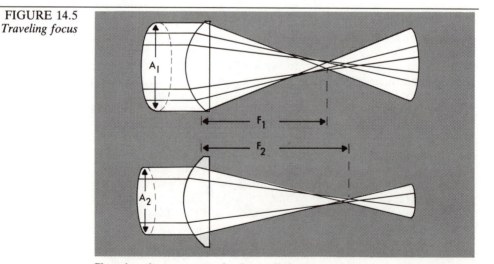

Changing the aperture of a lens affected by spherical aberration changes the point of best focus. Stopping down the aperture of such a lens may degrade sharpness instead of improving it.

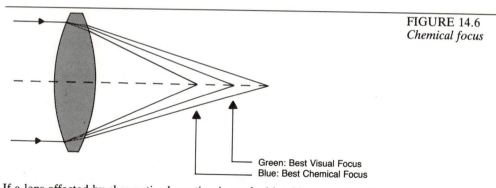

FIGURE 14.6
Chemical focus

Green: Best Visual Focus
Blue: Best Chemical Focus

If a lens affected by chromatic aberration is used with a blue-sensitive film, the point of best visual focus will be at variance with the point of best chemical focus. The image on film will be sharpest when blue rays are in focus. The eye focuses best on yellow-green rays. Unfortunately, when one is in focus, the other will be blurred.

Depth of Field

A special kind of image degradation comes into play when three-dimensional subjects are photographed and optically compressed into a two-dimensional plane. This problem has little to do with the quality of the lens; it occurs as a consequence of the limited depth of field of lenses.

A lens creates an image by causing light rays from subject points to converge and cross. For each minuscule subject feature the lens creates a cone of light that points toward the film plane. It is a cross section of this cone, known as the *circle of confusion,* that intersects the film plane and determines the size of the associated image point. See Figures 14.7 and 14.8. An image point will be rendered best, of course, when the film intersects the image cone at its tip where the cross section is narrowest. If the film intersects the cone some-where other than at the tip, image points will be enlarged and the image will be degraded accordingly.

If the subject is flat and parallel to the film plane, it will be rendered sharply on film when the tip of any one cone intersects the film plane. If the lens is a good one, all other points will come into focus on the same plane. If the subject is three-dimensional, image cones associated with subject features located outside the plane of focus will converge in front of or behind the film plane. It will be impossible to position the film so that it simultaneously intersects the tips of all image cones associated with a three-dimensional subject. This limitation is made clear by the lens formula:

$$\frac{1}{F} = \frac{1}{u} + \frac{1}{v}$$

where F is the focal length of the lens and v is the lens-to-image distance at which an object at distance u comes into focus. Since F

can be considered fixed for a specific image, object points in different planes must come into focus in different image planes.

Fortunately, image sharpness does not deteriorate entirely when the film is slightly displaced from the tip of an image cone; it merely tapers off. Image points whose rays converge close to the plane of critical focus are nearly as small as those at the critical plane. Objects close to the point of focus are thus rendered nearly as sharply as the object focused upon. A zone of reasonable sharpness extends in front of and behind the point of focus within which degradation is small enough to be tolerated. The distance from the front to the end of this zone is known as *depth of field*.

There is a corresponding zone at the film plane, known as the *depth of focus*, wherein slight errors in focusing can be tolerated. Within this zone, circles of confusion away from the plane of critical focus are almost as small as those at the critical plane. When the film is positioned within this zone, sharpness of focus is considered acceptable.

Relative sharpness in three-dimensional images can be improved by controlling image cones so that they form thin, narrow, pencil-like beams whose cross sections change gradually at increasing distances from the lens. The dimensions of the image cone vary first with the focal length of the lens. The size of the base of the image cone varies as well with the size of the aperture opening. The height of the image cone, the distance from the lens to the tip of the cone, varies with camera-to-subject distance. Short focal lengths, small apertures, and distant subjects form the narrowest beam and give the greatest depth of field.

From depth-of-field formulas it would appear that the image cone, the circle of confusion, and depth of field are not at all affected by

---

FIGURE 14.7
Circle of confusion

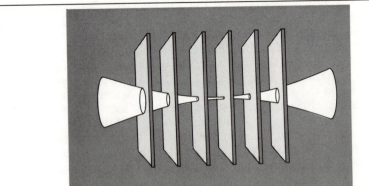

For each point on the subject, the lens projects a cone of light that converges toward the point of focus and then diverges. The diameter of an image point can vary depending on where the film intersects the cone associated with that point. The cross section of the cone where it intersects the film is called the circle of confusion.

the design or construction of a lens. Yet it is known that the confusion of rays at the point of focus is accounted for in part by lens-design errors, construction defects, and diffraction. These imaging errors do indeed alter the size and shape of the image cone. Traditional depth-of-field tables and formulas ignore these differences and assume that the image cone is geometrically exact, defined by straight lines and having a point-size tip. The depth-of-field formulas below make use of the same simplification; they greatly facilitate computations, but they do not accurately describe the way images are formed. Still, though inexact, they provide a useful approximation of true depth of field.

Depth of field is computed by finding D_f, the farthest distance where focusing is acceptable, and subtracting D_n, the nearest distance where focusing is acceptable. The near limit of depth of field, or the nearest distance at which the circle of confusion will be smaller than a given size, c, is given by:

$$D_n = \frac{u}{1 + \dfrac{un}{cF}}$$

The farthest distance at which the circle of confusion will be smaller than a given size is given by:

$$D_f = \frac{u}{1 - \dfrac{un}{cF}}$$

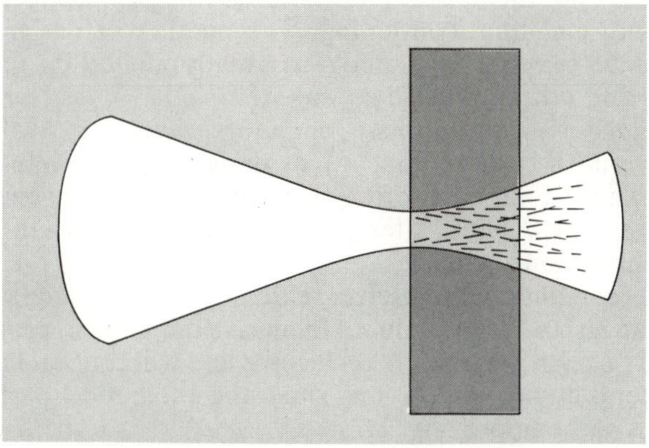

FIGURE 14.8
Circle of confusion

The cone of image light never comes to focus at a precise point. There is always some degree of confusion in the rays caused by lens errors or diffraction.

where

D_n = near limit of depth of field,
D_f = far limit of depth of field,
u = distance focused upon or camera-to-subject distance,
n = the f-number of the lens,
c = the diameter of the largest acceptable circle of
        confusion, and
F = the focal length of the lens.

The circle of confusion specified in optical formulas defines the amount of degradation acceptable in an image, or the amount of tolerable blurring. Even if you are unwilling to accept blurring, however, it is useless to compute depth of field using a circle of confusion smaller than the focal spot created by your lens at its point of best focus. You cannot, by defining an unreasonably small circle of confusion, use depth of field to improve image quality beyond the limits imposed by the quality of the system.

Improvements in Lens Design

There is much room for improvement in the designing of photographic lenses. Some long-hoped-for improvements are already on the way.

Computer-Designed Lenses

The theoretical performance of a lens is determined from its design by tracing the path of several rays through a mathematical model of the lens. A large number of complex calculations are required to trace just a few such rays, and the calculations must be repeated following each design change. Optical designers once made these ray calculations by hand using mechanical calculators. Because of the complexity of the optical formulas, the calculations were tedious, error prone, and time consuming. The number of design changes that could be feasibly evaluated was greatly limited by production schedules and other practical considerations.

Computers changed all this. Computers perform ray calculations so rapidly and accurately that a lens design can be evaluated with little delay and modified as often as desired. Before any lens is built, designers can now freely alter preliminary designs to see how theoretical performance changes.

Indeed, computers themselves can be programmed to initiate design changes and then evaluate them. A computer so programmed will modify design parameters on its own and will continue to change them as long as ray calculations show the theoretical performance of the lens to be improving.

Computer-generated designs have yielded lenses of superb quality, often produced at competitive costs. Indications are that the improvements have only just begun. The prospect of continued advancements in computer-aided design, along with the use of new optical materials, indicates that better lenses are definitely a part of photography's future.

More kinds of optical glass are now available to designers than ever before, some at relatively low cost. The high refraction and low dispersion of new optical glasses allow lens designers to combine elements in ways that yield superb correction of aberrations. An excellent example of such a new glass is crystalline calcium fluorite, which allows for superior correction of chromatic aberrations.

New Glass Types

You will recall that the chromatic aberrations of a lens, for the most part, are unaffected by the lens aperture setting and can only be reduced during the design of the lens. Until recently this was an incredibly costly task. Most general-purpose lenses manufactured before 1980 were *achromatic* at best. They were designed to bring light of two colors, red and blue, to a common focus. Such correction improves the performance of a lens considerably, but green fringing remains a problem, especially in long focal-length designs in which the error is magnified.

Color-Corrected Lenses

Apochromatic lenses, which are corrected in three colors, improve image sharpness substantially. Indeed, by correctly combining the newer glass types, designers have developed lenses corrected for more than three colors. Such lenses are called *superchromats*, and they are super sharp. Apochromatic lenses have been available and have been used in specialized optical applications for years, but they have not been widely produced for general photography until recently. Today, several apochromatic telephoto and zoom lenses are marketed for single-lens reflex cameras as well as for enlargers.

The spread function of a highly corrected lens at a given aperture may be so small that aberrations have less effect on image clarity than diffraction. Such lenses are said to be *diffraction limited*. When a lens is corrected in this way, it is as nearly perfect as a lens can become. See Figure 14.9.

Diffraction-Limited Lenses

Although the lenses used in general pictorial photography are seldom diffraction limited, diffraction-limited lenses are not impossible to construct or even hard to find. It is quite common, in fact, to find them used as telescope and microscope objectives. In these applications, a lens does not have to cover a wide field of view, so it is easy for the designer to reduce lens aberrations to an imperceptible level.

In contrast, lenses used in general photography must cover a substantial angular expanse and are more difficult to correct at the periphery of the field of view where the entry angle of the rays is fairly steep. It is quite impossible in fact to correct such lenses so that all aberrations fall below the diffraction limit. Apochromatic lenses come close to attaining this goal, but at large apertures and wide angles of view, coma and astigmatism, not the chromatic aberrations, are the really difficult problems to correct.

The superiority of highly corrected lenses is most apparent when they are used at large apertures with flat subjects. They are well suited as enlarger lenses where flat negatives are used to make flat prints. Luckily several lenses of apochromatic design and of very high resolving power are available for use with enlargers.

When diffraction-limited lenses are used with three-dimensional subjects, they have some of the same limitations as other lenses. It is impossible to obtain peak sharpness in an image over the entire depth of a three-dimensional subject regardless of the quality of the lens. A diffraction-limited lens used at wide aperture will bring a very shallow plane of a three-dimensional subject to critically sharp focus and will badly blur the rest of it. Used at a smaller aperture, it will create images vastly superior to those of a standard lens, but the margin of superiority will be reduced somewhat by the fact that imaging quality is reduced by diffraction at small apertures, even in a highly corrected lens. Thus while camera lenses of superb correction may be available for use with three-dimensional subjects, the return in image quality with respect to their cost has caused the less fortunate and less committed of photographers to forego purchasing them.

Despite the fact that a highly corrected lens may be expensive, if resolving power becomes all important, it will be worth the extra money. Such a lens is especially valuable to surveillance photogra-

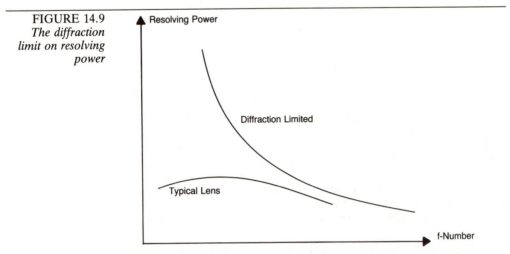

FIGURE 14.9
The diffraction limit on resolving power

Resolving power drops rapidly at small apertures or large f-numbers, even in diffraction-limited lenses. The resolving power of a general-purpose lens falls far below the diffraction limit at all apertures. There is much room for improvement in the design of camera lenses.

phers and to distance photographers. Distant subjects can often be rendered with adequate depth of field even when the lens is used near its maximum aperture, which in well-corrected lenses offers incredible sharpness.

When shopping for a lens, understand clearly what you want it to do. If you select the most expensive lens hoping that resolving power will rise proportionally with cost, you may make a bad choice. The relationship between imaging quality and the cost of a lens does not follow a predictable pattern. The features of some lenses, built-in shutters or automatic focusing, for example, increase their cost without improving their imaging properties at all. A highly corrected lens can be expensive, but a lens of lower resolving power may cost as much because it is two stops faster. Another lens may cost less yet excel over the others in imaging quality because it comes in a popular focal length and is manufactured in quantity.

Interpreting Lens Modulation-Transfer Functions

Deciding which lens is best for high-resolution work is simplified when one knows how to make sense of modulation-transfer functions. MTF curves may seem a bit strange at first, but do not be intimidated by them. They are explained very simply, and they give valuable information about the quality of a lens. More and more often, well-informed photographers are using modulation-transfer functions to evaluate lens quality.

MTFs indicate how well a lens holds contrast in the image, that is, how accurately the contrast of the subject is transmitted through the lens over a wide range of input frequencies. As mentioned in Chapter 3, frequency in the modulation-transfer function is related to the size of image features. Contrast is related to their visibility. Contrast ultimately determines whether features will be resolved. Contrast in the MTF is therefore the direct equivalent of visual quality. The lens whose MTF shows the greatest contrast, particularly at high frequencies, will yield the greatest clarity of image detail at those frequencies.

Comparing Lenses

Differences among good lenses can be so subtle that comparisons based on single MTFs give a limited, erroneous indication of how they compare. Lenses of high quality should be compared using a family of modulation-transfer curves. From a family of curves you may discover that the imaging quality of one lens is superior to that of another at one aperture setting, but inferior at a different setting. Or one lens may be designed to give maximum central sharpness, letting peripheral sharpness fall where it will, while another is designed to give an overall balance of sharpness with sharpness at the center sacrificed to improve performance at the edge. Only a family of modulation-transfer curves will provide the information needed to allow an informed evaluation of such lenses.

When using a family of curves, it is important to see that lenses are compared at the same aperture and that they are compared for rays at similar points in the field of view. Resolving power varies somewhat with lens aperture, and resolving power at the center of a lens may be double that at its edge.

The modulation-transfer function will not determine which imaging characteristics are important. It is up to you to establish your own criteria for favoring a lens. In the above example, is it better to select the lens with high central resolving power and low edge resolution, or the one with low central resolving power and even performance across its field?

The answer depends on how you intend to use the lens. In certain kinds of surveillance work where isolated details must be resolved, like a face or a license plate, the high central sharpness of the first lens may provide the extra resolving power needed to capture the image under difficult circumstances. In documentary or copy work where a flat field must be sharp from border to border, the lens with better overall sharpness may provide a more satisfactory image.

One approach to comparing the modulation-transfer functions of two lenses is to rank them by their frequency responses at a specific contrast level. Comparisons made at 20 percent contrast, the point where fine detail is just distinguishable, should allow one to rank lenses in the same relative order they would be ranked using resolving-power tests. In both cases the ranking is based on marginal resolution. Another approach is to compare lenses at a frequency where contrast drops below some optimum value, say 70 or 80 percent. These comparisons indicate the frequency range over which image clarity is nearly ideal.

A

B

PLATE 1
*Sharpness versus
resolving power*

C

These three prints were made from the same negative using different lenses. Lens A was a high-quality enlarging lens; lenses B and C were of less quality. The image made by lens C is apparently sharper than the one made by lens B, but lens B would rate higher than lens C in a resolving-power test. (From *Photographic Materials and Processes,* Stroebel, Compton, Current, and Zakia, courtesy of the authors.)

PLATE 2
Specular highlights

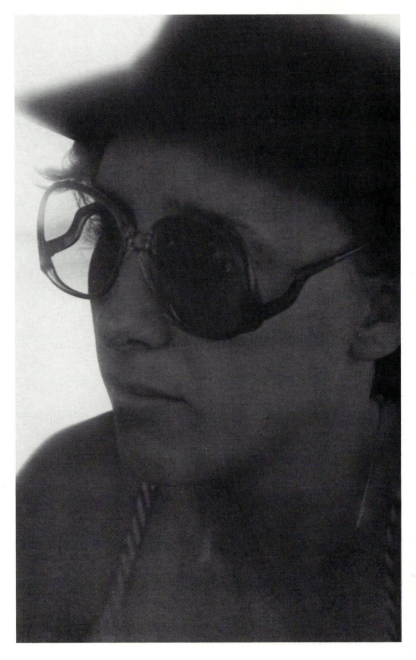

Specular highlights can make a crucial difference in one's perception of an image. The original image on the left, which has no specular highlights of consequence, is not only dull and lifeless but is softly focused near the center and badly out of focus elsewhere. The image on the right suggests how specular highlights (added by way of potassium ferricyanide bleach) give the impression of sharpness and brilliance.

PLATE 3
Diffuse lighting

These photographs demonstrate the capacity of diffuse lighting to conceal detail. The photograph on the right was made under tent lighting that caused all but the coarsest of textural details to weaken. The writing at the top of the figurine shows how dark the textural shadows would be if they were present; the writing proves that print exposure was not held back to achieve this effect.

PLATE 4
Tonality

Only three tones are present in the posterized image on the left: black, white, and a single shade of gray. The image on the right is the continuous-tone original. Despite its limited tonality, the three-toned image is a rather good representation of the original. In some respects it is better—its edges are sharper and its contrast is expanded—but the original has subtle tonalities and fine features that are not found in the posterization. These images illustrate two peculiarities about image reproduction: a finite number of tones can suggest the full tonal range, and one image can be sharper than another yet resolve less detail.

PLATE 5 Process films developed in gradation developers have extraordinary enlarge-
Enlargement ment latitude. This cropped photograph is a 13-power enlargement of the
latitude lower-left corner of the 35mm negative reproduced full frame in the inset.
The fineness of grain, the sharpness, and the tonal gradation in this print
are as good as would be expected from a large-format film.

PLATE 6
Solar flare

Solar flare creates multiple, ghost-like images of the diaphragm of the lens. As obvious as these flare patterns are, they are merely a special case of the general phenomenon of flare wherein each image point creates its own flare pattern. Because of the flare in this photograph, none of the shadows reproduces with the maximum black of the paper base, even though the print was made on a number-three contrast grade paper. The great danger of flare is that it lowers contrast and thereby lowers the resolution of detail.

PLATE 7
*Normal focal-
length lenses*

Although it may not be obvious from the fine-grain structure and the tonality of this photograph, the subject occupied only one-fourth the area of the negative and was cropped extensively in printing. This photograph exploits not only the generous enlargement latitude of a high-resolution film but also the impressive sharpness of a normal focal-length macro lens used at its optimum aperture.

PLATE 8
Wide-angle lenses

Wide-angle lenses give tremendous depth of field, even in scenes such as this, which was shot at a middle aperture (f/8 at 1/250th second). This photograph was made with a 20mm lens on 35mm process film developed in a gradation developer.

PLATE 9
*Depth-of-field
management*

When depth of field is inadequate to cover the full depth of the subject, it is more acceptable to preserve clarity of focus on the front of the subject at the expense of the rear.

PLATE 10
Peak of action

The nonstop action of the ballet leaves few opportunities for a shot free from blurring. At the peak of action, however, the subject is virtually motionless.

PLATE 11
Backlighting

These photos differ only in the direction of the light. Not only does back-lighting bring out better contrast and detail in the image on the right, but it eliminates the drabness that infects the frontally illuminated photo on the left. Backlighting makes the image sparkle.

PLATE 12
*Loading and
stressing*

It is in low-light situations like this that loading and stressing the tripod can make a difference. The reflecting pool shows that the wind was calm but active. To stress the tripod, a lightweight chain about six feet long was attached to the tripod head. By pulling the chain taut and standing on it to hold the stress, a steady tension was maintained throughout the exposure.

This close-up shot was made using a small amateur flash unit. This unit was not powerful, even as amateur units go, but at close range it produced enough light to allow the use of an aperture of f/22 with Kodachrome 64 film. Small flash units are ideal for this kind of work. The Kodachrome original was copied onto Kodak's Technical Pan Film to make the black-and-white print.

PLATE 13
Close-up flash

PLATE 14
Jumbo rocks
Thirty-five millimeter photographers can get detail and tonal gradation in their photographs without upgrading to large-format equipment. This photograph, as well as most others in this book, was made on 35mm film—Kodak Technical Pan Film 2415 developed 15 minutes in a modified POTA developer (see Formula 17.6 in Chapter 17).

Investigators and researchers are relying increasingly on compact, portable, small-format cameras. High-resolution techniques are essential for success in their work.

PLATE 15
Fossil

PLATE 16
T-grain emulsion

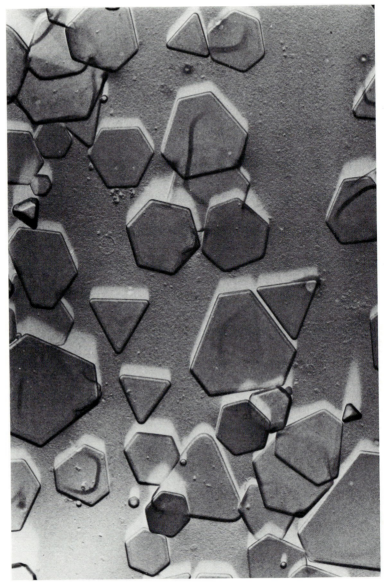

T-Grain emulsion

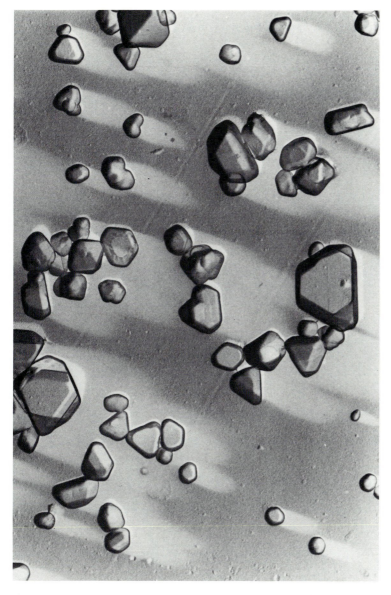

conventional grain emulsion

The crystals in T-grain emulsions lay flat so that their largest projective surfaces face the imaging light. This orientation gives the emulsion some remarkable properties. (Reprinted courtesy of Eastman Kodak Company.)

These photographs were made late in the day when the sun was low and shadows were long. The top left photograph, which was exposed according to the f/16 exposure rule, has dense shadows. The bottom left photograph, exposed a full f-stop more or at one-half the exposure index of the first, shows shadow detail more as it appeared to the eye. The photograph above, exposed two f-stops more or at one-fourth the exposure index of the first, has better detail in shadows as compensation for the sacrifice in effective film speed. Each negative, being on roll film, was developed identically. Had sheet film been used, the negative used to make the top left photograph would have benefited from increased development, which would have brightened the highlights. The negative used to make the photograph above would have benefited from curtailed development, which would have prevented highlights from washing out.

PLATE 17
Film speed versus shadow detail

PLATE 18
Image streaking from subject motion

The streaks on the left are characteristic of linear subject motion. A shutter speed of 1/500th second, adequate to freeze ordinary movement, could not freeze these high-velocity water droplets.

Chapter 15

Image Displacement
Sources of Image Motion

Image displacement refers to a change in the position of an image during exposure due to movement of either the subject or camera while the shutter is open. Many photographers learn belatedly that a minuscule disturbance can visibly reduce the level of detail resolved on film. Spread components measured in microns, too small to be sensed by sight or touch, can be sensed by high-resolution systems. Various physical vibrations encountered in photography introduce blur components measuring hundreds, even thousands, of microns—enough to neutralize the photographer's most careful preparations.

Some photographers underestimate the damage that can be done by image motion because of the widely held notion that blurring will not reduce clarity if the blur component is smaller than the point-resolution threshold of the eye. This is incorrect. Blurring from image motion combines with other sources of image spreading to increase degradation cumulatively. Even if all spread components are smaller than the threshold of visibility, they will each combine so that, unless each is considerably smaller, the cumulative-spread function may easily exceed the point-resolution threshold.

Furthermore, photographers sometimes incorrectly estimate the effect of an increase in shutter speed. For example, doubling the shutter speed will not halve the spread function as is sometimes implied; it merely halves the motion component of the spread function. See Figure 15.1.

The magnitude of the blurring from image motion depends on the velocity of the image at the film plane and the span of time the motion acts on the film during the exposure:

$s_x = vt$

where s_x is the spread component created by the motion, v is the speed of the image at the film plane, and t is exposure time. If the image moves as slowly as 10 millimeters per second during an exposure of 1/500th second, a blur component of 20 microns will be added to the spread function.

When image displacement is linear, the limiting resolving power R_m of the system cannot exceed:

$$R_m = \frac{1}{s_x}.$$

In the example just cited, resolving power will never exceed 50 lines per millimeter at a shutter speed of 1/500th second.

Camera Motion

Common sources of camera motion are ground vibration, body motion, and camera vibration.

Ground Vibrations

You may feel confident about the solidity of the ground beneath you, but many massive structures taken as solid are not as stable and motionless as they appear to be. It is estimated that the ground and floors in cities vibrate with a magnitude of about 2 to 10 microns at a rate of about 10 to 20 vibrations per second; the tops of modern skyscrapers can sway in a stiff breeze by several feet; and a bridge constructed of concrete and steel may swing several inches from side to side on a calm day.

Ground vibration is affected by such uncertain phenomena as wind, automobile traffic, the operation of heavy equipment, and

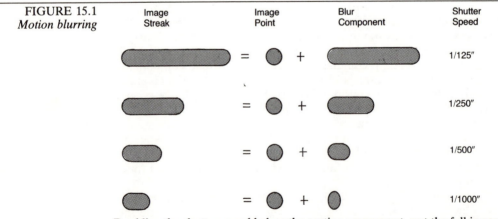

FIGURE 15.1
Motion blurring

Image Streak		Image Point		Blur Component	Shutter Speed
	=	◯	+		1/125″
	=	◯	+		1/250″
	=	◯	+		1/500″
	=	◯	+		1/1000″

Doubling the shutter speed halves the motion component, not the full image streak of motion blurring. At a fast enough shutter speed, motion blurring will be reduced to a manageable level. Still, even at very fast shutter speeds, motion blurring may continue to reduce image clarity.

local mechanical disturbances. It is a variable phenomenon whose severity is difficult to predict. It is troublesome because the magnitude of the motion it causes is often too small to be sensed; it may go unnoticed until a negative is processed. It is of less concern at fast shutter speeds, which offer the best defense against the nuisance. During time exposures longer than about a quarter second, however, the full effect of ground vibration can register on film.

It is virtually impossible to hand hold a camera and keep it still. **Body Motions** Even if you hold your breath, your body stays in constant motion from the action of your pulse. Indeed, no matter how well you brace yourself, you cannot escape the pulse's periodic and perpetual jerking motion. The magnitude of the motion imparted to a hand-held camera by the body varies from one individual to another, but the motion imparted by the pulse alone can be as great as 0.2 millimeter, or 200 microns, with the jolt lasting about a tenth of a second.

When exposure time is so long that pulse vibrations are fully transferred to the image, blurring can be quite pronounced. Beyond that, the specific effect pulse vibrations may have on the quality of hand-held photographs is unpredictable. At fast shutter speeds it may happen that an exposure begins and ends entirely between heartbeats. Consequently, at shutter speeds substantially faster than a tenth of a second pulse displacement may or may not occur at the instant the shutter is open. Yet even at very fast shutter speeds, there is a significant risk in hand-held photography that the image will be degraded by body motions of one kind or another, either from breathing, the pulse, or the minuscule muscular motions that act continuously to stabilize the body's position.

Microvibrations in single-lens reflex cameras are created by the **Internal** camera's own internal mechanisms, primarily by its shutter and **Camera** reflex mirror. Any noise made by a camera is a certain sign of **Vibrations** vibration. The louder the noise, the greater the magnitude of the vibration. Single-lens reflex cameras with their moving mirrors and focal-plane shutters emit a loud characteristic clatter betraying their complex internal mechanisms.

Mirror Vibrations.   When the shutter-release button of a single-lens reflex camera is depressed, spring tension moves the reflex mirror out of the image-forming path so that light transmitted by the lens can reach the film. As the mirror moves, it quickly builds speed until it crashes into the damping material at the top of its swing. The momentum of the mirror is then transferred to the body of the camera according to the relationship

$$w_1 v_1 = w_2 v_2$$

where w_1 and v_1 are the weight and velocity of the mirror, w_2 is the weight of the camera, and v_2 is the force imparted to the camera by the mirror. Some of the force from the mirror is absorbed or damped by spongy material in the camera's body. The remainder is transferred to the camera's body as a vibrational motion.

The reverberation, which can be significant, can be felt by the photographer and under certain conditions can actually be seen. For example, the force of the mirror is great enough to propel a free-spinning turntable over a distance of several inches. The concussion can displace the camera enough to produce double images on film. In less severe cases, the displacement shows up as a general streaking in the direction in which the mirror travels.

Shutter Vibrations.   Any vibration caused by the shutter is of concern because it occurs at the instant of exposure. Rangefinder and twin-lens reflex cameras are nearly vibration free because they usually have leaf shutters and have no moving mirrors. The operation of a leaf shutter is such that the moving blades of the shutter pivot away from the center so that the motion of one blade is opposed by the motion of a blade on the opposite side. The shutter blades are thereby self-damping. In addition, the moving parts of these shutters are usually less massive than those of focal-plane shutters and they operate with less force. Their quietness is evidence of their smoothness.

Subject Motion

The blurring patterns created by subject motion are usually less uniform than those of camera motion. If one part of the subject moves faster than another, it will blur more. Those parts of the image that are motionless reproduce more sharply. Also, the closer the subject is to the camera and the more parallel its motion to the film plane, the worse it blurs. Blurring is lessened in distant subjects that move in the direction of the lens axis. In effect, blurring is affected more by the angular speed than by the absolute speed of the subject.

When a simple motion occurs along a straight line at a constant speed, image displacement is easily analyzed. Figure 15.2 illustrates a motion that occurs in the direction of the large arrow at an angle a to the film plane. The distance m indicates how far the subject moves during a given time interval and is related to the speed of the motion. The focal length of the lens is given by F, and the distance from the lens to the subject is given by u. The displacement d in the image that results from a motion m is proportional to the angular displacement of the subject m'. The apparent motion m', not the absolute motion m, gives the relative speed of the subject and determines the magnitude of image blurring. Although the two motions are different, they are visually similar. Were it not for a gradual change in the size of an object as it moves away from or

toward the camera, one would be hard pressed to distinguish between absolute and apparent motion.

For distant subjects the triangle defined by the sides m and m'
and by the angle a can be approximated by a right triangle. Apparent
motion is therefore approximately equal to

$m' = m(\cos a)$.

From the lens formula the magnitude of the motion at the film
plane is found to be

$$d = \frac{m'Ft}{(u - F)}$$

Replacing m' by $m(\cos a)$ gives

$$d = \frac{mF(\cos a)t}{(u - F)}$$

which can be simplified using the relationship

$$M = \frac{F}{(u - F)}$$

where M is image magnification. The factors that determine the
magnitude of blurring from subject motion are made clearer in the
approximation

$d = Mtm (\cos a)$.

Blurring increases in direct proportion to an increase in image
magnification, exposure time, subject speed, and the cosine of the
angle at which motion occurs.

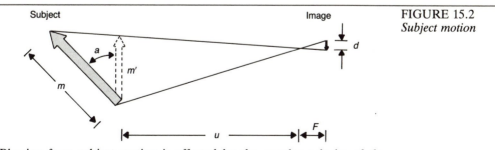

Subject   Image   FIGURE 15.2
                                                                Subject motion

Blurring from subject motion is affected by the angular velocity of the
motion m', camera-to-subject distance u, lens focal length F, and exposure
time.

More about this equation will be found in Chapter 20. In the meantime, the equation allows an interesting comparison to be made of the damage done by camera motion to that done by subject motion. Whenever the subject is reproduced smaller than life-size, subject motion generates less blurring than will be generated by an equal motion at the camera position. Consider, for example, a situation where a moving subject is photographed from 5 meters away. If the subject moves in a direction parallel to the film plane so that the effect of its motion is fully recorded, the angle of motion with respect to the film plane will be 0 and its cosine will be 1. Image magnification will be approximately 0.01 if a 50mm lens is used. Blurring from this motion at a given shutter speed will be 100 times smaller than if the camera moves the same amount. Camera motion in this case will cause more damage to image clarity than the equivalent subject motion.

The comparative damage, however, depends largely on image magnification. The situation reverses when the subject is reproduced greater than life-size. Then subject motion generates greater blurring. Close-up photography, already complicated by various exposure problems, will be complicated further if the subject is in motion.

Part III

High-Resolution Techniques

From Part I you know that peak resolving power cannot be achieved until each source of degradation is identified and controlled. When reaching this goal is impossible, you know the importance of identifying and acting on major sources of degradation. You know from the discussion in Part II that there are many ways photographic clarity can be lost. Here in Part III you will learn specific techniques that make use of the principles discussed in Part I to manage the mechanisms of clarity and degradation discussed in Part II. It is time to put your camera to use in practicing and validating these techniques.

Chapter 16

Exposure
Burning an Image onto Film

It is clear that precision in exposing photographic films aids in preserving image clarity. In this chapter you will see what is meant by correct exposure, and you will learn what to look for in your negatives to tell whether or not your method of exposure is correct. You will also learn about exposure meters and discover why these handy devices cannot always be trusted.

Film Speed

To expose film expertly, you must know the exposure index that works best with your film-developer combination, which may or may not be the film's ISO speed rating. The ISO film-speed rating is a useful guide that leads to accurate exposure for most purposes, provided your meter is correctly calibrated and correctly used. Even so, you may ultimately decide, for good reason, to use some other speed rating as your working exposure index.

Development and Film Speed

One reason you may wish to customize your exposure index is that the effective speed of a film depends somewhat on the extent of development. From a family of characteristics curves, as in Figure 16.1, you will see that as the degree of development increases, the film's threshold exposure shifts to lower log exposure values, an indication that the film will respond in a useful way to a lower level of light.

Official Film Speeds

The official speed rating of a film is determined by a specific test in which a film is exposed to a log luminance scale of 1.3 and developed to a density scale of 0.8. Film speed is based on the camera exposure that gives a shadow density in the negative of 0.1 above the base-plus-fog density. Regrettably, the official rating pro-

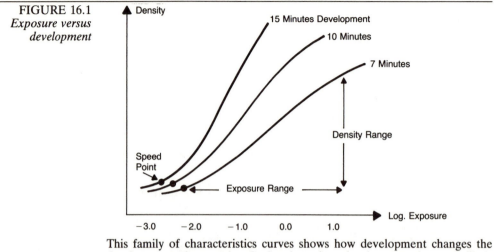

This family of characteristics curves shows how development changes the density range of a negative. It also shows how development changes a film's exposure index. Lengthening development time shifts the film's speed point to the left, improving shadow speed.

cedure requires a greater degree of development and gives a higher contrast index than is commonly sought in pictorial photography. The increase in development moves the speed point of the emulsion to the left on the characteristics curve and yields a more optimistic speed rating than will be achieved in ordinary pictorial development.

Chapter 18 contains details of a test you may conduct to find an exposure index precisely tailored to your working methods. For the time being, since the exposure index you settle on as a result of your test is likely to be close to the ISO speed rating, you can count on getting usable images in all but the most critical situations using the ISO film-speed rating.

Evaluating Exposure

It is clear enough that exposure must be precise. To make certain that it is you must know how to determine whether or not your film has been correctly exposed. In making this determination you must evaluate two properties of your negatives: shadow detail and overall density.

Shadow Detail

Ideally, exposure through the thinnest areas of a negative must always be great enough to produce solid black in a print. This requirement suggests that print-exposure time can never be less than a certain minimum. A necessary corollary is that density in important image areas in a negative must be great enough for these areas to hold detail and contrast when the negative is given a great enough exposure to produce the maximum practical print density of your print paper. Adequate shadow tonality in the negative, and consequently in the print, can be achieved only when film exposure, specifically the exposure associated with important shadow luminances, is greater than the threshold exposure of the emulsion. After all, this is the only way the associated image areas can build density in the negative. This principle is vitally important to obtaining correct exposure.

It is exposure more than development that guarantees the negative will be sufficiently dense in shadow areas to hold visible detail and contrast when the thinnest areas of the negative are printed for maximum print density. If shadow areas are exposed on film below the threshold exposure of the emulsion, it will be impossible to force usable density in these areas, even with extended development. This is the reasoning behind the old maxim: expose for shadows, develop for highlights.

Shadow density, more than highlight density or the density of midtones, is used to evaluate exposure. You need not be concerned about the density in midtones and highlights when density is sufficient in shadows. Also, do not be alarmed at this point if a correct shadow exposure causes excessive density in highlights. If it does, it merely means that the degree of development of the film was too great and must be adjusted.

Overall Density

The second consideration in determining whether a film was correctly exposed is the overall density of the negative. The thinnest negative that yields full print tonality including maximum black is the preferred negative. This rule must be applied relative to the previous rule. That is, one must never compromise shadow density for the sake of making thin negatives. The point is simply that the negative should be as dense as needed to obtain good shadow detail and no denser.

Negatives should not be excessively dense for several reasons. First, the image in a thin negative will form nearer the surface of the emulsion and will consequently be sharper (see Chapter 11). Also, the graininess that stems from energetic development in heavily exposed areas will be less likely to occur (see Chapter 12). Print exposures will be shorter as well, which makes for an easier time in the darkroom. Yet more important is that thin negatives are made at smaller apertures that improve depth of field or faster shutter speeds that reduce motion blurring. Film exposure should therefore be set to obtain the thinnest negative in which important subject detail records above threshold.

Under-exposure

An underexposed negative has too little density in shadow areas. When such a negative is printed to obtain maximum black in deep shadows, important shaded areas that should contain texture are too thin and also print black. If print exposure is reduced in an attempt to improve shadow tonalities, the print will have no deep black tones at all and shadows will be weak and unconvincing. The presence of weak shadow values and the absence of strong black tones in a print are likely signs that the film was underexposed.

Overexposure

A film has been overexposed when density in the thinnest parts of the negative is substantially greater than that of the film base. Areas in a negative that represent deep shadows, those permitted to reproduce as solid black in the print, should be just perceptibly denser than the base-plus-fog density of the negative. Based on measurements taken from high-quality negatives, density in the deepest shadow areas need not be more than about 0.1 units above base-plus-fog density. If it is more, exposure was greater than necessary and the film has technically been overexposed.

Overexposure can also be defined from the perspective of highlight density. A film has been overexposed when highlights in the negative record on or beyond the shoulder region of the characteristics curve. When negatives are this dense, tonal separation in important print highlights will be reduced; highlights will be chalky, low in contrast, and devoid of detail. It is in this respect that thick-emulsion films have so much latitude for overexposure. Indeed, they can tolerate overexposure of 10 or more f-stops while retaining tonal separation in highlights. They are not easily overexposed.

Deliberate Overexposure. Many professional photographers deliberately and routinely overexpose negative materials, both black-and-white and color. They use exposure indices smaller than the suggested ISO film speed rating partly because ISO film speeds are often higher by the equivalent of one-third to one-half f-stop than experience indicates they should be.

Other professionals prefer generous exposure as a safety margin against an occasional error that leads to underexposure. They reason that any loss of resolution caused by slight overexposure is less important than the more catastrophic, total loss of shadow detail that would result from underexposure. Overexposure captures the image; underexposure loses it. Neither optimizes resolving power, but having an overexposed, printable negative certainly makes better sense than leaving an assignment with underexposed, blank film.

Correct
Exposure

Exposure is pragmatically correct whenever it produces a negative that yields a full-gradation print: a print having the deepest black of the paper emulsion, the whitest white of the paper base, and a wide variety of tones in between. Exposure is optimally correct when it yields the thinnest negative capable of giving a full-gradation print.

An optimally exposed negative is one having a density reading in the thinnest image area of about 0.1 units above the film's base-plus-fog density. Density in the thinnest image area of an overexposed negative will be much greater than 0.1 units above base-plus-fog, and density in the thinnest image area of an underexposed negative will be as thin as in the unexposed film borders.

If you want to evaluate your negatives for correct exposure and do not have a densitometer, do not despair. By performing a simple experiment you can identify a correctly exposed negative without one. Begin by making a series of exposures on your favorite film varying exposure in increments of one-third f-stop. Begin the series several f-stops above the exposure indicated by a meter as correct and continue for several f-stops below it. Make certain that the frame receiving the least exposure is decidedly underexposed and the frame receiving the greatest exposure is decidedly overexposed. Thus you will be sure that one image in the group is perfectly exposed or so near to it that the difference will not matter.

You now have only to find the optimally exposed negative. Look first at the underexposed negatives to determine if density in the thinnest image areas differs from density in the clear nonimage area between frames and near socket holes. The first negative to show a perceptible density above base-plus-fog in the thinnest image area is optimally exposed or close to it. This evaluation will be easier if you frame the subject so that a deep shadow falls at the edge of the frame where it can be compared side-by-side with a clear area in the margin of the film.

The outcome of this test will be more accurate if you conduct it after you have determined the correct development time for your film-developer combination as explained in the next chapter. Otherwise, if your development time changes by much, your exposure index may change as well. Still, even if you find subsequently that your film's development time must be modified, the exposure index you establish now should be correct within a fraction of an f-stop assuming you have made no errors in running the experiment.

Exposing Photographic Films

The luminance of the image that strikes the film is controlled by the aperture setting of the lens. The length of time this light acts on the film is controlled by the camera's shutter speed. The aperture and shutter speed dials on modern cameras are versatile devices that increase or decrease the level and volume of light arriving at the film plane so that virtually any subject luminance value can be made to reach the film at any desired meter-candle-second (MCS) exposure level.

The general idea is to adjust these exposure controls so that the volume of light arriving at the film plane matches the exposure range of the film. In other words, adjust MCS exposure so that it coincides with the portion of the film's sensitivity range falling between the toe and shoulder on the characteristics curve. As shown below, there are several ways to determine the camera settings that accomplish this end.

Exposure Meters

Most photographers rely on exposure meters to find their aperture and shutter-speed settings. Exposure meters measure either the light reflected from the subject or the light incident upon it.

Reflected-Light Meters.   To appreciate the subtlety of operation of the reflected-light meter, you should understand the distinction between a middle and an average luminance value. If you do not, you are certain to have occasional exposure problems when you use one. A *middle luminance* is the luminance associated with a gray object whose reflectance is midway between white and black. A diffuse reflectance of 18 percent is widely accepted as a standard middle reflectance. The *average luminance* of a scene will be reasonably close to the middle luminance, provided that the proportion of light and dark tones in the scene are balanced. If a single light or dark tone dominates the metered frame, it will cause the average luminance to shift away from the middle value toward the value of the dominant tone.

Reflected-light meters measure the average luminance, but are calibrated on the assumption that a middle luminance value will be metered. This is the fallacy in their operation. They indicate the exposure that will reproduce the metered luminance with a middle density in the negative so that it creates a middle gray in the print.

Consider how a reflected-light meter reacts in a situation in which a deep black object and a bright white object are photographed together. When these objects appear in a correctly exposed and processed negative, the black object will record with a density just greater than the base-plus-fog density of the film. The white object will record with a density well up on the straightline portion of the film's characteristics curve. Assuming both to be of equal size, their average luminance will be close to that of a middle luminance and a reflected-light meter will indicate the exposure setting needed to create such a negative.

What happens if the white object is now replaced with a black one so that two black objects appear together? The average luminance of the subject and the level of light reaching the meter will be lower and the meter reading will change. The exposure that renders one black object correctly in the first scene, however, is the exposure needed to render two black objects correctly in the second. One black object, or two or three, must be exposed to record on film with a density just above base-plus-fog density—and no greater. The meter indicates the wrong camera settings in the second situation. Should you use the indicated exposure, it would render the black objects with the density of gray objects and would render all other print tones proportionately lighter.

The exposure setting indicated on the dial of a normally calibrated meter causes the metered brightness value to record with a medium negative density, typically of about 0.65 to 0.70 above base-plus-fog density or roughly midway up on the straightline portion of the characteristics curve. If the metered area consists of two widely separated luminance values, the indicated exposure has the effect of centering the corresponding negative values above and below 0.65 to 0.70 on the characteristics curve. If the scene is dominated by a single luminance value, the indicated reading will cause that value to record at 0.65 to 0.70 density on the curve. In the latter case, the indicated exposure will be correct only if the metered tone happens to be a middle value.

To their credit, reflected-light meters are usually quite accurate, especially when bright areas and dim areas occupy equal parts of the image so that their average luminance is close to a middle value. When subject luminance values are balanced in this way, readings from reflected-light meters can usually be trusted. When the luminance of the subject changes because its reflectance value has changed or when a scene contains more bright areas than dim areas, it is risky to follow blindly the exposure indicated by a reflected-light meter.

Exposure must be based more on the level of light striking the subject than on the reflectance of the subject. This is a principle worth remembering. Under conditions of equal illumination, the correct exposure for a black subject will be very nearly the correct exposure for a white or gray subject.

Some photographers, those who think all their negatives should have consistent density, have a difficult time accepting this premise. Exposing for consistent density is a misguided objective. Low-keyed subjects, those having a preponderance of dark tones, print best from thin negatives. Dark objects, whether one or many, should reproduce in the negative with density below a middle density. Likewise, high-keyed subjects having a preponderance of bright tones print best from dense negatives.

Shadow Exposure Readings.   It should be clear by now that film exposure is optimally correct when important shadow areas are exposed just above the film's threshold exposure. Therefore, the most accurate way to use a reflected-light meter is to measure shadow luminance. First the photographer must decide which shaded areas are important enough to be exposed above threshold and which others are less important and may be allowed to print black without detail. Not only is this decision a matter of taste, but if the photographer is to predict correctly how exposure decisions will render print tones, he or she must have an intimate working knowledge of the exposure characteristics of the film being used. It is simpler, particularly for inexperienced photographers, to identify the area that should reproduce as a middle tone, somewhere between maximum white and maximum black. It is also simpler for manufacturers to design meters so that they read the average brightness at the scene.

If you are determined to do so, you can use an averaging meter to make a shadow exposure reading, but you must then adjust the indicated camera settings. First, identify and meter the area you want to record as textured black, not deep black, in the print. An example might be dark, sunlighted foliage. The camera settings indicated on the dial will, of course, render the metered area as middle gray instead of black. To get the expected tonality, reduce the indicated exposure by three full f-stops. This reduction will cause the exposure of the metered area to fall just above the film's threshold exposure on the toe of the characteristics curve. The metered area will subsequently be rendered in a normal print as dark gray with discernible textural detail. Reducing the indicated settings by more than three stops will cause the exposure of the metered area to fall below the film's threshold. In this case, the metered area will reproduce blank in the negative and black in the print.

Incident-Light Meters.   Incident-light meters measure illumination from the light source: light incident upon, instead of reflected from, the subject. Variations in subject reflectance have little bearing on the exposure indicated by an incident-light meter, so exposure is usually more reliable. In the situation discussed earlier, wherein a white object was photographed along with a black one, then two black objects were photographed together, an incident-light meter would have correctly indicated that identical exposures be made.

Incident-light meters do not always lead to perfect exposure, however. When a scene is filled with numerous bright objects or dominated by a bright background, reflected light can raise ambient illumination so that shaded areas are flooded with light and lighting contrast is reduced. On a beach, in a desert, or in snow, for example, light reflected by large expanses of white scenery weakens textural contrast and creates chalky highlights. Such a scene is often rendered better when given less exposure than indicated by an incident-light reading.

At the opposite extreme, when many dark-toned objects dominate a scene, there will be little ambient reflection and textural contrast will again be low. In this case, it is better to expose the film more than indicated by an incident reading so that negative density is raised above the toe of the characteristics curve for improved shadow separation.

Exposure Tables

Exposure tables are quite reliable guides when used under the exact lighting conditions specified in the table. Their accuracy stems from the principle that exposure depends on the nature of lighting at the scene more than on subject reflectance. Tables are generally interchangeable among films of equal ISO speed ratings.

Sunlight Exposure

Exposure in frontal sunlight can be determined accurately and consistently using the f/16 exposure rule. This rule lets one determine the correct exposure for any film from its ISO speed provided the subject is illuminated frontally in direct sunlight. By this rule, the shutter speed in direct sunlight at a lens aperture of f/16 is the reciprocal of the film's ISO speed rating. It is not necessary, of course, to actually use f/16

during the exposure; having found the basic settings, you may substitute any equivalent combination. If the ISO rating of the film is
32, you may use 1/15th second at f/22, 1/30 second at f/16, 1/60th
second at f/11, and so forth.

A few precautions should be heeded when using this rule. First,
it applies to bright sunlight—not overcast daylight, twilight, or dawn.
The sun must be bright and high in the sky if exposure is to be
accurate and consistent. Second, it applies to frontal sunlight. Those
parts of the subject struck directly by sunlight will be correctly
exposed, those parts shaded will be underexposed.

Not many rules in photography work equally well in every situation.
There are certainly exceptions to the rules given here for exposure.

Exceptions to the Exposure Rules

Close-Up Exposure.   Light meters and exposure tables assume that
the lens-to-subject distance is infinite. If it were so, light rays arriving
at the lens from the subject would all be parallel. When the subject
is closer than infinity, as it always is in practice, rays from image
points diverge on their way to the lens, reducing image brightness
in proportion to their divergence. Image brightness at the film plane
therefore varies with the lens-to-subject distance. The closer the
subject, the greater the divergence of its rays and the dimmer the
image.

At lens-to-subject distances greater than a meter or two, the
difference in illuminance from that at infinity is just a fraction of an
f-stop, small enough to be disregarded. At reproduction ratios approaching and exceeding unity, the loss becomes more significant
and requires compensation. For example, at a reproduction ratio of
1:4, the loss is about half a stop. At a 1:2 ratio, it has reached a
full stop, and at 1:1, two full stops. Through-the-lens exposure
meters read this loss directly and compensate for it automatically.
When close-up exposure is determined by a detached meter, one
must compensate manually by increasing the indicated exposure.

Failure of Reciprocity.   The rehalogenation of silver atoms at the
sensitivity speck discussed in Chapter 11 is the basis for what is
called the low-level failure of the reciprocity law. The reciprocity
law is summarized in the equation:

$$H = ET$$

where H is exposure, E is the illuminance of the exposing light as
determined by the aperture, and T is the duration of exposure as
determined by the shutter speed. This relationship accounts for the
fact that various combinations of aperture and shutter speed yield
identical exposure. If the level of the exposing light is halved by
closing the aperture one stop, an equivalent exposure can be arrived
at by doubling the time the exposing light is allowed to act on the

film. As the level of light striking the film is reduced, however, eventually so few photons arrive at individual crystals that stable latent-image centers do not form in time to prevent single atoms of silver from reverting to ions and rehalogenating.

It is not the mathematical relationship that breaks down during the so-called failure of the law. That is, it is not the accumulation of luminous energy that changes during rehalogenation. The failure, if it must be called that, is with photographic emulsions, which are not consistent at all exposure levels in their response to light. Emulsion crystals do not absorb light or retain it with the same efficiency at all exposure levels. When the level of the exposing light is extremely low or extremely great, the action of a photon of light can be reversed within the emulsion. Exposure, which indeed accumulates according to the law, does not always translate into an equivalent increase in silver density. The reciprocity law predicts how much light will arrive at the film, not how the film will respond to it.

During time exposures, in which the level of the exposing light is low, exposure must be greater than that indicated by a meter to compensate for this so-called failure of the law. The exact compensation needed depends on the film and on the shutter speed used. For example, see Table 16.1. For many films, exposures longer than a second or two will require compensation of a full stop or more. Consult reciprocity information included with the film or obtained from the manufacturer to determine the exact compensation needed.

Unusual Lighting.   Unusual lighting, such as numerous specular reflections in a scene, make it difficult to use a meter accurately. Specular reflections are essentially mirror reflections of the light source that occur when light striking the object is fully reflected. One sees such

TABLE 16.1   Reciprocity characteristics differ for different films. If you use exposures
Reciprocity   longer than 1 second or shorter than 1/10,000th second, you will need to
adjustments   compensate for reciprocity failure by increasing exposure.

| If Indicated Exposure Time Is (seconds) | Kodak T-MAX 100 Professional Film | | | Kodak T-MAX 400 Professional Film | | |
	Use This Lens-Aperture Adjustment	OR	This Exposure-Time Adjustment (seconds)	Use This Lens-Aperture Adjustment	OR	This Exposure-Time Adjustment (seconds)
1/10,000	+ 1/3 stop		Change aperture	None		None
1/1000	None		None	None		None
1/100	None		None	None		None
1/10	None		None	None		None
1	+ 1/3 stop		Change aperture	+ 1/3 stop		Change aperture
10	+ 1/2 stop		15	+ 1/2 stop		15
100	+ 1 stop		200	+1 1/2 stops		300

Reprinted courtesy of Eastman Kodak Company.

reflections often on wet or glossy surfaces. These reflections, only slightly dimmed by reflection from the surface, are very bright and have a significant influence on exposure meters when they appear in an image in quantity. Solving the exposure problem they create can be difficult even if you have a spot meter and can measure small areas of the scene directly. If you expose to render the diffuse surfaces accurately, the bright specular reflections can cause halation. If you expose to tone down specular reflections, you may underexpose diffuse tones.

Unless you have previously solved the problem posed by such tricky lighting situations, the surest way to arrive at correct exposure may be to bracket your exposures. Bracketing means making several exposures, one at the metered setting and others above and below it in increments of one f-stop or a fraction of an f-stop. Bracketing improves the odds that one of the images will be successful. Bracketing also provides, in separate frames, detail in brightly lighted as well as in dimly shaded image areas. In many scientific and technical applications it is important that information be recorded in both highlights and shadows, even if not on the same piece of film. Bracketing gets this information.

Bracketing

Bracketing can be enlightening as well. By observing how the image changes at various exposure levels, one can learn to predict how a fixed difference in exposure will affect image tonality and clarity. In this way, bracketing can teach you something about the tonal characteristics and exposure latitude of the films you use and it may ultimately improve your skill in exposing these films accurately.

Chapter 17

High-Resolution Development
Finding the Miracle Developer

Developers promoted as high-resolution formulations are explored in this chapter. The chemical basis of certain wonder formulas—fine-grain developers, acutance developers, speed-enhancing developers, and gradation developers—are examined to see how they work and to see whether they perform as claimed. Fine-grain developers have caused considerable consternation among experimental photographers. In view of the proliferation of developers claiming to be of this genre, a discussion of fine-grain developers is a good way to begin.

Fine-Grain Developers

Fine-grain development is generally achieved by solvent development, physical development, or a combination thereof.

Solvent Fine-Grain Development

Developer solutions that contain silver-halide solvents reduce graininess by reducing the size of emulsion crystals during development. An image speck thus grows from a smaller crystal, its size being reduced according to the power of the solvent action. Solvents also encourage physical development, which as will be seen below, further promotes fine-grain development. Commonly used solvents are sodium thiocyanate, ammonium chloride, and sodium thiosulfate. See Formulas 17.1 and 17.3. These powerful chemicals are capable of dissolving silver halides entirely if given the opportunity. Sodium thiosulfate, for example, is hypo, the ingredient in fixers that removes undeveloped emulsion crystals. In concentrated form it will clear a film in a matter of minutes, removing all traces of silver halide from an emulsion.

Metol	5	grams
Sodium sulfite	100	grams
Kodalk	2	grams
Sodium thiocyanate	1	gram
Potassium bromide	0.5	gram
Water to make	1	liter

FORMULA 17.1
Solvent fine-grain developer

Kodak's DK-20 developer is a slow-working fine-grain developer. Sodium thiocyanate is used to liberate silver ions from emulsion crystals as fuel for physical development. Sodium thiocyanate is a powerful silver solvent.

Although solvent fine-grain developers do reduce the size of image specks and reduce the grainy appearance of the image, they provide no corresponding improvement in resolving power. By all expectations, an increase in resolving power would be accompanied by an increase in information capacity. This is hardly accomplished by solvents. Indeed, depending on their concentration in the developer solution, they can entirely dissolve the smallest emulsion crystals, destroying information. As a result, films developed in solvent developers lose some of their sharpness, detail, and gradation.

Solvent development is also accompanied by a loss of emulsion shadow speed. A film's exposure index can drop by as much as two f-stops, one-fourth its rated speed, when processed in a solvent developer. The exposure index of an ISO 125-speed film processed in the solvent developer DK-20 may drop to 32 or less. As a rule, image quality will be better when an ISO 32-speed film is used and processed normally. The slower film processed in its recommended developer will likely be less grainy than the ISO 125-speed film processed in a solvent fine-grain developer, and it will almost certainly give greater sharpness and resolving power.

Sulfite Fine-Grain Development

Sodium sulfite is a weak silver-halide solvent that dissolves emulsion crystals slightly whenever it is used in a developer, which is almost always. Sodium sulfite, you may recall, is the antioxidant or preservative most frequently used in photographic developers. As an antioxidant its concentration in a developer never has to be greater than about 30 grams per liter. Still it often appears in fine-grain developers at two to three times this concentration. Developers like D-76 and D-23, for example, contain 100 grams per liter of sodium sulfite. See Formulas 17.2 and 12.2. Even at such high concentrations, however, sodium sulfite is weaker in its solvent action than some of the powerful solvents discussed above.

The enduring popularity of D-76 and D-23 among distinguished photographers suggests that these sulfite fine-grain developers may be of more than passing value and may represent a difference in kind from the caustic-solvent developers examined earlier. Apparently the solvent action of sodium sulfite is strong enough to counteract the growth associated with filamentary development, but weak

FORMULA 17.2 *Sulfite fine-grain developer*	Metol	2 grams
	Sodium sulfite	100 grams
	Hydroquinone	5 grams
	Borax	2 grams
	Water to make	1 liter

This fine-grain developer, designated D-76 by Kodak and ID-11 by Ilford, has enjoyed continued popularity since its introduction in 1929. It gives tight grain and good emulsion speed.

enough to preserve image information; it reduces grain with less reduction in resolving power.

Solution Physical Development

Whenever silver ions and solid silver appear together in solution, the ions are drawn toward and attached to the solid silver as though being electrochemically plated. So another consequence of having solvents in a developer solution is that they release silver ions into the solution to fuel this plating process.

$$AgBr \rightarrow Ag^+ + Br^-$$

The solid silver to which the ions plate themselves are the silver specks of the latent image center or the silver filaments at development sites. Because this plating action builds density roughly in proportion to the size of the image speck, or in proportion to exposure, the process increases the density of a negative in the same proportion as does ordinary development. It is therefore considered a kind of development itself. It is called *physical development* to distinguish it from chemical or direct development. See Formula 17.3.

The silver specks formed by physical development are dense and compact rather than loose and filamentary. Also, images produced by physical development tend to form where latent image centers are largest: near the surface of the emulsion. Physical development thus favors those crystals least affected by light scatter. Consequently, images obtained by physical development are characterized by extremely fine grain and excellent sharpness.

Pure physical developers, formulated to maximize physical development while minimizing chemical development, are of experimental interest, but because of their low emulsion speed, are of little practical use. Before physical development can be induced, a fairly large seed of solid silver must be present, much larger than that needed to initiate chemical development. Exposure must thus be comparatively massive in films that are to undergo physical development so that large latent image centers are created from the effect of exposure alone.

FORMULA 17.3
Physical developer

PART A		
Sodium sulfite	50	grams
Silver nitrate	32	grams
Sodium thiosulfate	150	grams
Water to make	1	liter
PART B		
Metol	2	grams
Sodium sulfite	10	grams
Hydroquinone	3.4	grams
Sodium hydroxide	3.4	grams
Water to make	1	liter

The special components in this physical developer are: sodium thiosulfate, a powerful silver-halide solvent that dissolves emulsion crystals and brings their silver ions into solution; silver nitrate, which introduces additional silver ions into the solution; and sodium hydroxide, a pwerful accelerator that speeds chemical development so that the size of the silver seeds of the latent image centers are increased before the emulsion crystals are completely dissolved.

Combined Physical and Chemical Development

When chemical developing agents are added to physical developers, the latent image center builds from chemical development and the exposure index of the film rises dramatically toward its inherent speed rating. Developers of this kind usually contain a powerful solvent and a high-energy accelerator. The accelerator hastens the activity of the chemical developer so that the latent image seed builds somewhat from chemical development before solvents in the solution completely dissolve the crystal. At the same time, the powerful solvent dissolves the crystal before chemical development proceeds too far.

The conditions necessary to initiate physical development are straightforward: silver ions must be present in the solution along with a seed of solid silver. These conditions occur whenever an exposed film is introduced into a solution containing silver ions or a silver-halide solvent. In practice, physical development almost always occurs during film development since nearly all developers contain the solvent sodium sulfite as a preservative.

As a rule, physical development, even as a secondary process, counteracts chemical spreading and reduces graininess. When chemical and physical development occur simultaneously, the nature of the image's grain structure will depend on the extent to which physical development predominates. The longer physical development acts on the silver specks, the denser the filaments of the image specks become. Thus one can formulate a large variety of fine-grain developers by modifying standard chemical developers to increase the ratio of physical development to chemical development.

One way to increase the role of physical development is to increase the concentration of silver ions in the developer solution. Silver ions may be supplied in large quantity either by adding silver nitrate to the solution or by adding a solvent to dissolve the silver-halide crystals of the emulsion. The latter option accounts for part of the fine-grain effect given by the solvent and sulfite developers discussed earlier.

Physical development operates slowly compared with chemical development and is therefore encouraged by long developing times. Lengthening the time required for complete development, provided this does not increase the final contrast index, reduces grain. When a film develops quickly, chemical development brings the image to full density before physical development can proceed very far. Lengthening the time needed for full chemical development, say by using dilute developers, allows physical development to operate longer and exert a greater effect on image characteristics. Also, if development time is increased without increasing the contrast index, it means that the energy of development will have been reduced. The lower energy of development will further inhibit graininess.

Dilute Developers.   Dilute developers lengthen developing time, increase the role of physical development, and reduce graininess. In doing so, however, they weaken shadow density. The reason is that the small amount of developing agent available in the dilute solution is consumed in dense areas of the negative that develop early. By the time shadow regions begin to build density at the end of the developing cycle, the developing agent is virtually exhausted. This loss of shadow detail can be prevented by giving the film more exposure. In this way shadow density will build earlier in the development cycle before the developing agent is consumed. As noted before, however, increasing the exposure level is tantamount to reducing the exposure index. Consequently, dilute developers improve grain at the expense of emulsion speed.

Low-Alkalinity Developers.   The energy of development can be reduced without reducing the concentration of the developing agent by replacing the alkali with one of lower alkalinity. A low-alkalinity accelerator simply slows down the chemical reaction. Toward the end of the development cycle, a plentiful supply of developing agent remains in the solution to act in low-exposure regions. Meanwhile, physical development operates longer to decrease grain. Fine-grain developers formulated with low-alkalinity accelerators do not lower emulsion speed as dilute developers do. The improvement they bring to image quality is significant as it is achieved without sacrificing other image characteristics.

Modern Fine-Grain Developers

Modern fine-grain developers are based on a combination of rapid-induction developing agents, low-alkalinity accelerators, and superadditive combinations. Typically, the rapid-induction developing agents Phenidone or Metol are used in superadditive combination with hydroquinone in a mildly alkaline solution. An antioxidant is added to improve the useful life of the solution. Sodium sulfite, the antioxidant used in most developer formulas, is often present in greater concentrations than needed for its antioxidant role. An organic antifoggant like benzotriazole, which provides better fog protection than an inorganic bromide and retains emulsion speed better, is added in a small amount.

From this outline one can formulate a very good fine-grain developer by working out the concentrations of the various chemicals. The appropriate combination and concentration of ingredients will depend largely on how the developer will be used—on such considerations, for example, as production volumes and whether the solution will be replenished or will be used as a one-shot developer.

The most famous and enduring published fine-grain formula is the MQ developer introduced in 1929 by Capstaff, called D-76 by Kodak and ID-11 by Ilford. The formula has been in continuous use since its introduction. Many professional photographers use it today as their standard developer. Used with modern high-speed emulsions, D-76 achieves very good emulsion speed with moderate grain. Its large concentration of sodium sulfite, 100 grams as opposed to the 25 or 30 grams needed to preserve the developing agent, provides a moderate solvent action that releases silver ions into the developer solution and increases the contribution of physical development. The solvent action also etches the surface of crystals to reveal buried latent image centers and provide additional points of attack for the developer solution.

Phenidone developers and developers formulated from Phenidone derivatives have properties superior to those of MQ developers, and these properties translate into improved image quality. Many proprietary fine-grain developers are in fact based on Phenidone or its derivatives and are quite good. Recent experiments in fine-grain development have been devoted to finding new applications for Phenidone and its derivatives.

Conclusions about Fine-Grain Development

Graininess is affected by three emulsion properties: the size of individual image specks, the clustering of these specks in the emulsion, and the random distribution of specks and clusters. Only one of these properties, the size of individual silver specks, can be altered during development. Since either increasing or reducing the size of image specks can cause an undesirable reduction in image information, the specks are best left at their inherent size.

It was once thought that clustering of the silver-halide crystals occurred during development and that this clustering was the principal effect to be countered by fine-grain developers. This was more correct several decades ago than it is now. Graininess increases and clustering accelerates in untreated, raw emulsion preparations whenever the emulsion is wet and softened. Modern emulsions are hardened and treated with restrainers to prevent continued crystal growth following ripening. There are indications that clustering during development in modern emulsions is virtually nonexistent. Except for the filamentary growth of silver or the explosive extrusions of silver that can occur during high-energy development, development seems hardly to induce the wholesale migration of crystals, or anything resembling it. Clustering occurs instead when the film is made. If the graininess of modern films increases at all during development, it seems to occur for other reasons.

Certain developers emphasize grain so much that their operation seems to refute this statement. Speed-enhancing developers, for example, produce negatives with pronounced grain in which clustering appears to have worsened. Even so, it is unlikely that graininess worsens due to the migration of halide crystals during development. Speed-enhancing formulations often achieve their effect by forcing the image to form in the largest, most sensitive crystals. The smaller crystals occupy space in the emulsion but are not a part of the final image. Consequently the larger crystals are surrounded by additional blank space so that the contrast and visibility of image grain increases. Also, many of these developers require an extended development time which forces the image to form deep in the emulsion where light scatter is greater and image points are larger.

One should not be unduly concerned about the graininess of photographic films. Merely accept the fact that it is an inherent property of photographs due to the nature of the silver-halide process. Furthermore, keep in mind that graininess does its harm when the negative is enlarged. If graininess seems unacceptable at a particular print size, it means that the enlargement latitude of the negative material has been exceeded. The solution is to switch to a film with the required enlargement latitude or keep enlargements within the latitude of the film. When you have no choice but to use a high-speed, coarse-grain emulsion, accept the fact that graininess will be pronounced.

Acutance Developers

Developers formulated to improve image sharpness are known as acutance developers. Acutance developers achieve their effect either by means of surface development or adjacency effects.

Surface Development

Surface development exploits the fact that rapid-induction developers operate quickly on heavily exposed crystals. Since the crystals at the surface of an emulsion are exposed more heavily than crystals below the surface, surface crystals are acted upon by rapid-induction developers earlier in the development process. Rapid-induction developers thus form images close to the surface where crystals have been affected less by light scatter. As a result, the damage done by turbidity is reduced and sharpness is improved.

A surface developer can be formulated from a weak solution of Metol activated by a highly alkaline accelerator. Such a developer is characterized by rapid development in heavily exposed regions and a loss of potency as development proceeds. Both characteristics are necessary to the process.

The improved sharpness achieved by surface development will be lost if the negative is fully developed. When development is carried to completion, deeper parts of the emulsion layer build density and the surface effect diminishes. Herein lies the weakness of surface developers. The reduction in the degree of development needed to heighten the surface effect weakens negative contrast. To bring highlights to a normal density, exposure must be increased. Since this necessary increase in exposure is equivalent to a reduction in the exposure index, surface development achieves its effect at the expense of emulsion speed.

Nearly all modern developers are formulated using rapid-induction developers. They thereby routinely promote surface development and can be considered acutance developers provided the film does not get developed to excessive density. When a film is developed normally in a modern developer to obtain optimum gradation and a moderate contrast index, acutance will also be optimized. Overdevelopment lessens image sharpness because it forces the image to form deep in the emulsion where light scatter is greatest.

Adjacency Development

Adjacency development, although it provides an improvement in apparent sharpness, provides no genuine increase in acutance and is not in common use today. The value of discussing it here is in the insight it provides into the chemistry of development.

Adjacency development depends for its effect on the way the activity of Metol is restrained when soluble bromides are present in a Metol-based developer solution. Bromide ions get into a developer solution in two ways: (1) they are added deliberately, often in the form of potassium bromide, to improve developer selectivity; and (2) they are released as a by-product of development whenever an emulsion contains silver-bromide crystals. This second mechanism, the release of bromide ions during development, is exploited in adjacency developers. Since all popular films contain silver bromide as the primary silver halide in the emulsion layer, all are capable of adjacency development.

When molecules of developer react chemically with a silver-bromide crystal, they produce atoms of silver—the primary product of the photographic reaction—and ions of bromide as a by-product. These bromide ions originate at and concentrate around development sites, the only locations where they can exercise a restraining effect on the developing crystals. If these pockets of bromide are not dispersed into the solution by periodic agitation of the developer, their restraining effect will increase to the point where development is curtailed significantly. This is why agitation is such an important part of normal development. Adjacency developers, on the other hand, depend on this bromide accumulation around development sites.

When a heavily exposed area appears in an image next to an area of lower exposure, the heavily exposed area, upon development, will contain a higher concentration of bromide than will the lightly exposed area. At the edge where the two areas meet, excess bromide from the high-exposure area disperses laterally. As it spreads into the low-exposure region, it reduces development all along the edge. A thin line is thereby formed whose density is lower than density in the center of the low-exposure region. This line is called the Mackie line or the fringe effect. See Figure 17.1.

While this is happening, a high concentration of unused developer remains in the low-exposure area because the developer solution is not used as heavily there as in the adjacent high-exposure area. This fresh developer diffuses laterally into the high-exposure area so that, at the edge on the high-density side, a thin line forms along which density is increased. This is known as the border effect. The combined outcome is that the two areas are separated along their borders by lines of higher contrast than the areas themselves. Visual separation is improved and edge definition is enhanced.

Developers that oxidize rapidly encourage adjacency development. The developer should therefore be used diluted. The concen-

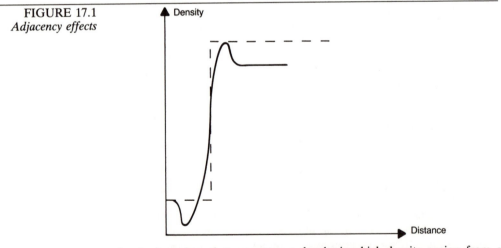

FIGURE 17.1
Adjacency effects

At the boundary that separates a developing high-density region from a low-density region, accumulated bromides can diffuse into the low-density region and restrain development in a way that produces a line whose density is lower than that in other parts of the low-density region. At the same time, unused developer from the low-density side diffuses across the boundary to create a line of higher density on the high-density side. The result is exaggerated contrast at the edge that enhances apparent sharpness.

tration of sodium sulfite should be reduced as well to speed the exhaustion of the developer. The effect is further enhanced by keeping agitation to a minimum to avoid dispersing the bromide buildup. The film must lay flat and must stay still during development. Agitation to remove air bubbles should be restricted to the first few seconds of development. Also, development cannot be allowed to proceed to completion or the effect will be lost. Upon full development, the low-density Mackie line and the restrained high-exposure area will both develop up to their normal densities.

The Beutler acutance developer is the best known of the adjacency developers. It is similar to D-23 except that it has a lower concentration of both Metol and sodium sulfite. See Formula 17.4. The low sulfite concentration allows rapid oxidation. To further speed oxidation, Metol can be kept to concentrations as low as one gram per liter.

Adjacency developers are necessarily Metol based. Phenidone is not restrained significantly by bromides. Hydroquinone is affected only by large quantities of bromide.

Speed-Enhancing Developers

Any process that increases emulsion speed improves resolving power by allowing the use of a faster shutter speed or smaller aperture. A genuine increase in emulsion speed, however, is one obtained without sacrificing another film quality. From this perspective, increases in film speed are technological breakthroughs not to be taken lightly. Many speed-enhancing developers are far from breakthroughs. They sacrifice shadow detail, grain, or some other film quality in achieving the increase.

Push Processing

Some commercial speed-enhancing formulas involve little more than high-energy, rapid-induction developers and push-processing techniques. Simply stated, push processing entails extended development that forces a film to a higher contrast index. It can be seen from characteristics curves that the speed point and therefore the exposure index of a film changes with development. At their highest contrast

FORMULA 17.4
Beutler acutance developer

PART A	
Metol	5 grams
Sodium sulfite	25 grams
Water to make	1 liter
PART B	
Sodium carbonate	25 grams
Water to make	1 liter

The Beutler developer, the best-known acutance formula, exploits adjacency effects. The key to obtaining improved edge sharpness with this developer is to keep film agitation to a minimum.

index, however, the real increase in the shadow speed of most films is less than one full f-stop. When rich, radiant shadow detail is needed, the usable increase in speed is generally closer to about a third of a stop.

Films that are to be push processed are often rated at exposure indices 2 to 4 times their ISO rated speed. It is not uncommon to hear of speed ratings 6 to 8 times the ISO speed. At such inflated exposure indices, the increase in speed is not genuine because detail and the sensation of luminance in shadow areas is lost. The effect of push processing here is merely to force marginally exposed highlights and middle values up to a printable density.

Push processing a film which has been mistakenly underexposed is a useful emergency technique. When the technique is properly executed, a film underexposed by two or three f-stops can be processed to capture an image in situations where the image would otherwise have been lost. No matter how well the technique is executed, however, the increase in the exposure index is achieved with a loss in shadow contrast. This is a natural consequence of underexposure. When a film is underexposed by more than a fraction of an f-stop, dim shadow areas register below the exposure threshold of the emulsion. Extended processing cannot compensate for this deficiency. Beyond achieving a small improvement in the speed point, extended development can do nothing more in areas receiving subthreshold exposure than increase fog density.

Rapid-Induction Developers

The induction speed of the developer has an important bearing on the exposure index of the film. Rapid-induction developing agents and other ingredients that improve developer induction speed are consequently of great value in high-resolution photography. This useful property helps to explain the popularity of the rapid-induction developing agents Metol and Phenidone, and the frequent use of sodium sulfite and hydroquinone in modern developer solutions. Sodium sulfite speeds the induction of development and thereby gives a useful improvement in a film's exposure index. Combining hydroquinone with Metol or Phenidone creates a superadditive combination that further improves induction speed and exposure index (see Chapter 12).

Gradation Developers for High-Resolution Films

The slow, fine-grain films that are esteemed in high-resolution photography benefit little from fine-grain or acutance development. These films, having inherent fine grain but tending to develop to a high density range, benefit more from gradation development.

Films intended for general pictorial use must be capable of producing a complete gray scale in the print, that is, a wide range of tones extending from the deepest black to the brightest white. Although the structure of the film determines whether it can be gradation processed, the developer determines whether gradation will in fact be achieved. The challenge to high-resolution photographers is to obtain gradation development using films that offer the highest resolving power. As it happens, the finer the inherent grain and the slower the speed of a film, the greater the challenge becomes.

The difficulty of developing high-resolution films for tonal gradation arises because the resolving power of a film and its tonal gradation are each determined by opposing consequences of the same emulsion parameter. Emulsions with small crystals (high resolving power) automatically have a narrow range of crystal sizes (low gradation).

It appears that process films, the high-contrast films used in graphic arts and in photomechanical reproduction, are at a transition point between films that can only be developed for high contrast and those that are gradation developable. Recent activity in gradation development has thus been directed toward improving the response of process film to gradation treatment.

Dilute Developers

One approach to obtaining gradation from high-contrast films is to develop them in highly diluted low-energy MQ developers. The effect of diluting the developer in this case is to lower the contrast index of development as well as to encourage physical development. Some success, although not much, has been achieved with such developers. A limitation of diluted developers was discussed earlier in the section on physical development: they are characterized by a loss of developer activity during late stages of development that reduces effective film speed and weakens shadow detail.

POTA Developers

Important progress in the formulation of high-resolution gradation developers came with the introduction of POTA-type developers. See Formula 17.5. POTA is a Phenidone-sulfite developer prepared without an accelerator or restrainer and without hydroquinone for superadditive combination. When Phenidone is used alone with general-purpose films exposed to normal-contrast scenes, it produces negatives of such low contrast that they are unprintable. Only when the contrast of the subjects is extraordinarily high, on the order of a million-to-one, does POTA produce a useful density range in general-purpose films.

The most promising use of POTA in high-resolution photography has been in obtaining tonal gradation from high-contrast, fine-grain

---

Sodium sulfite	30	grams
Phenidone	1.5	grams
Water to make	1	liter

FORMULA 17.5
POTA wide-latitude developer

POTA is a low-energy, wide-latitude developer used with coarse-grain films to tame subjects of wide luminance scale and used with fine-grain films to obtain gradation development.

---

process films. As a bonus, POTA and its derivative formulas yield higher exposure indices with these films than do dilute MQ developers. See Formulas 17.6 and 17.7. The Phenidone-sulfite developer brings out good detail in low-exposure areas without sacrificing middle values and highlights. Because of the fine-grain characteristics of the film itself, image detail and tonal gradation are superb. Even at exhibition-sized enlargement ratios, graininess is low. Of even greater significance is the increase in resolving power achievable with these films compared with only moderately faster wide-latitude films. Their resolving power is so great that virtually every source of degradation must be eliminated from the image before the full potential of these films can be realized.

Conclusions about High-Resolution Development

Studies have shown repeatedly that the effect of development on the imaging characteristics of a film is small compared to the inherent differences among films themselves.

Resolving Power

Despite improvements in the grainy appearance of a photograph brought about by certain fine-grain developers, a film's resolving power cannot be improved during processing, whether by reducing emulsion crystal size or by increasing it. To increase resolving power, one must add to the film's information capacity, a feat that cannot be achieved during film processing.

Exposure Index

Certain fine-grain and acutance developers lower the effective speed of a film. Caustic solvent developers and pure physical developers, for example, require massive exposure. One would do well to avoid a developer that reduces film speed unless an associated improvement in image quality warrants it. Likewise, one may wish to avoid a developer that increases speed at the expense of image quality. Film speed is determined largely by inherent characteristics of the emulsion as indicated by the speed point and threshold exposure on

FORMULA 17.6
POTA derivative

Sodium sulfite	25	grams
Phenidone	1.4	grams
Benzotriazole (0.2%)	15	milliliters
Borax	0.8	gram
Water to make	1	liter

This modified POTA developer, designated Delagi #8 after its originator Dick Delagi, a senior scientist with the Texas Instruments Company, is formulated to yield gradation development in process films like Kodak's Technical Pan Film. It gives results similar to that given by POTA, but with less fogging.

Sodium sulfite	85	grams
Hydroquinone	5	grams
Borax	7	grams
Boric acid	2	grams
Phenidone	0.13	grams
Potassium bromide	1	gram
Water to make	1	liter

FORMULA 17.7
PQ fine-grain developer

This is Ilford's ID-68 formula, one of the earliest published Phenidone formulas. Other PQ formulas can be derived by modifying Metol-based formulas.

the film's characteristics curve. It is indeed possible to obtain usable images from a film using inflated exposure indices, but such images are obtained at a cost. An acceptable developer has only to yield the full emulsion speed of a film without weakening shadow detail or worsening grain.

The improvement achieved by fine-grain development is often cosmetic. That is, it increases the visual appeal of the image and lowers the visibility of grain, but does nothing to improve image detail and gradation. The resolution, sharpness, and tonal gradation of an image are best ensured by the use of a fine-grain, high-resolution film.

Graininess

The chemistry of high-resolution photography is greatly simplified when high-resolution films are used from the start. Many of the objectionable properties of coarse-grain, high-speed emulsions simply do not occur in fine-grain, thin-emulsion films and they cannot be made to occur accidentally during processing. Fine-grained, thin-emulsion films do not exhibit excessive chemical spreading even with vigorous development, and they are less turbid than faster films even when overexposed. When high-resolution films are used in combination with gradation developers, ordinary care in film processing is sufficient to bring out the peak resolving power of the film; special processing techniques are unnecessary.

Chapter 18

Tone Control
Coordinating Exposure, Development, and Printing

The fact that the print paper, the medium in which the photographic process is consummated, has a fixed maximum reflectance ratio imposes a special limit on the process of image formation. The actual tonal relationships of a subject can be accurately displayed in a print only when the paper's maximum reflectance ratio equals or exceeds the subject's luminance scale. When it does not, tonal relationships must be compressed somewhere in the reproductive process so that the brightest and dimmest subject tones can be represented within the highest and lowest reflectance values of the print medium.

Print tonality can be altered in so many ways, either during exposure, development, or printing, that photographers are often at a loss to discover how to control it best. The issue is complicated further when image clarity and resolving power are considerations. By mastering a system of tone control—some systematic approach to film exposure, film development, and printing—the photographer can bring order to these processes so that certain procedural changes become predictable and image contrast and gradation can be altered at will. Without order, tone control can be perplexing.

Refined systems of tone control, such as the zone system of Ansel Adams, seek to capture the mood of the subject and to evoke in the print the sensation of luminance experienced at the scene. Achieving this effect is as much art as science. The goal of tone control in high-resolution photography is to enhance information in the image and to preserve clarity. Clarity and mood in a photograph are by no means mutually exclusive, but the techniques used to recreate the atmosphere of a scene may not be the ones that maximize clarity.

If all subjects were photographed under similar lighting, achieving optimal image tonality would be a trivial problem; but photographers deal with subjects of different reflectance value illuminated by light of varying contrast using many different films, papers, developers, developing tanks, and processing procedures.

The presence of so many variables and options makes the exchange of information about tone control seem futile. Serious photographers therefore establish personalized systems of tone control based on the results of tests conducted with their equipment and their processing procedures. In establishing a system of tone control, one must establish three baseline standards: print-exposure time, film-development time, and film-speed index.

Tone-Control Baseline Standards

The standard print-exposure time is the shortest print exposure that produces deep black from a negative area of minimum density, as in the unexposed margins of the film, and concurrently produces deep black from a negative area whose density is 0.1 units above the film's base-plus-fog density. If a negative prints well at this standard exposure, one can be confident that it has been properly exposed and properly developed. Negatives that print well only at a shorter exposure time are either underexposed or underdeveloped and will give weak black tones as a result. Negatives that print well only at longer exposure times have been overexposed or overdeveloped.

To determine your standard print-exposure time, use a negative that has a clear area (equal to the base-plus-fog density) adjacent to an area whose density is about 0.1 units greater (just perceptibly greater) than the base-plus-fog density. Set the aperture of your enlarger lens to a middle position, say f/8, and make a series of prints at exposures that differ by a few seconds, making sure that both areas are printed in each exposure. Try to produce a graduated test strip in which the longest exposure yields deep black and the shortest exposure does not.

Examine the prints in this series beginning at the dark end and find the last print in which the clear area and the adjacent 0.1 density area both print at the same density. This density is the maximum useful print density of the paper—about 90 percent of the paper's absolute maximum density. The exposure used to make this print will be your standard print exposure. It is the shortest exposure you can ever use in straight printing to produce a print containing deep black from the thinnest areas of your negatives.

Since you will use this printing exposure in future tests, make sure you can duplicate it exactly. A simple, but inflexible, way to do so is to make all test prints at the same enlargement ratio using

Standard Print-Exposure Time

the same aperture and exposure time. Another way is to meter a clear area in the projected image of the negative using a darkroom exposure meter, then adjust the aperture of the enlarger lens to get the same illuminance at the baseboard during subsequent tests.

Standard Film-Development Time

The *optimum* development time for a film is that which creates a negative of ideal contrast and gradation such as described in Chapter 10. The optimum development time varies from scene to scene, however, as subject luminance range changes. The *standard* film-development time is that associated with a scene of some fixed luminance range. This scene may be whatever you choose, perhaps something typical of what you photograph. A frontally sunlit outdoor scene is suggested as a standard subject if you have no other preference. Such a scene is considered to have a normal luminance range. If you live in a sunny region, you would do well to perform this test outdoors on a clear, sunny day using a subject with a range of reflectance values from black to white. Position the camera during the test so that the sun strikes the subject from the front or within a 45-degree angle of the front.

By conducting this test in frontal sunlight, you can determine correct exposure using an exposure table or using the f/16 exposure rule discussed in Chapter 16. Make a series of identical exposures on the film you wish to test. In the darkroom, cut the film into strips containing about five frames each and develop the first strip at the manufacturer's recommended development time. Make a print from one of the negatives using your standard minimum print-exposure time and evaluate the results.

If all has gone well, the shadow areas of the print will be deep and rich with detail. If they are not, you have probably made an error in exposing the film or in determining your standard print-exposure time. If you have made an error, correct it and repeat this test until shadows have depth of tone and visible detail.

From a correctly exposed negative, you can evaluate your development time based on the rendition of print highlights. If highlights are chalky white with little tonal separation, film-development time was too long and must be reduced. Develop another strip of the test roll at 70 percent of the previous time to see whether highlights improve. If highlights are muddy and dark, film-development time was too short and must be increased. Continue to fine-tune film-developing time until a print of perfect tonality is made at your standard print-exposure time. When such a print is made, the development time used for that negative will be your standard film-developing time.

Film-Exposure Index

As a result of changing your development time in the preceding test, you may need to adjust your exposure index somewhat. You can fine-tune your exposure index using the test that follows.

Using a frontally sunlit subject, make a series of exposures varying each in increments of one-third f-stop beginning several f-stops above and ending several f-stops below the exposure given in an exposure table as correct. Develop the film at the standard time found in the previous test. The correctly exposed negative in the series will be the first to show a density of 0.1 units above the base-plus-fog density of the film in an area of the scene representing deep shadows. Refer to the procedure demonstrated in Chapter 16 if you have no way to measure film density. Determine your exposure index from the selected negative by applying the f/16 exposure rule in reverse. That is, translate the aperture and shutter speed used to make the selected frame into an equivalent exposure at f/16. Your exposure index will be the reciprocal of the equivalent f/16 shutter speed.

Although the negative selected in this way is the correctly exposed negative, some photographers disregard the objective criterion and base their exposure index on a subjective evaluation of shadow detail in their prints. Their reasoning, as you will discover during these tests, is that one trades film speed and sharpness for shadow detail, and vice versa. Thin negatives, being made at a faster shutter speed or a smaller aperture, are essentially exposed at a higher exposure index. From the favorable camera settings used to make them and from their reduced turbidity at the lower level of exposure, the thinner negatives should be somewhat sharper than the correctly exposed ones. The denser negatives, on the other hand, give prints with better tonality and brilliance and with greater clarity in shadow areas. Should you decide to sacrifice shadow detail to increase your exposure index or to sacrifice effective film speed to improve depth of tone, you may adjust your exposure index as you see fit. (See Plate 17.)

Contrast and Gradation Controls

Having determined your three baseline tone-control standards to your satisfaction, you may use any of several techniques to alter contrast and gradation to fit subject tones into the limited contrast range of photographic papers. Your first option is the deliberate sacrifice of tonality at one or both extremes of the luminance scale. In practice, photographers simplify the problem of a contrasty scene, such as a scene in bright sunlight, and narrow the effective contrast range by allowing certain shaded areas, which may contain contrast and detail clearly visible to the eye, to reproduce in prints as deep black. They also allow certain bright features, light sources and specular reflections for example, to reproduce with the maximum white of the paper base.

Rather than attempt to reproduce in a print the absolute luminance range of a contrasty scene, from the brightest specular highlights to the darkest shadows, it is more reasonable to concentrate on those highlights and shadows that must print with texture. When this approach is taken, the relevant luminance scale of contrasty sunlight typically drops to about 160. Since this modified scale exceeds the reflection range of print papers, however, it will still be necessary to compress image tones so that they reproduce within the paper's narrower reflectance range.

Photographic contrast can be altered at the negative or print stage by using any of the eight methods given below. Unfortunately, having too many ways of arriving at what is ostensibly the same image tonality can overwhelm as much as it can help. Practical systems of tone control treat all but one or two of the available contrast controls as constants by always using them in the same manner or by not using them at all.

Film Development

Customized film development is a flexible and predictable way to modify contrast and gradation. It is fundamental to tone-control systems like the zone system wherein negatives are developed individually at some multiple or fraction of the standard developing time, depending on subject luminance range. The idea in the zone system is to raise or lower the density range of each negative by an amount that allows each to print perfectly on a middle-grade print paper. This procedure is best suited to large-format systems that take sheet film. In 35mm and other formats that take roll films, it is usually more practical to develop the roll as a unit. Roll-film users are more likely to standardize their film-development time and to depend on other contrast controls to customize image tonality.

Lighting Contrast

Elaborate systems of tone control are less necessary in a controlled setting, such as in a studio, where uniform lighting allows film and print processing to be standardized. In situations of predictable or repeatable lighting, the perpetual experimenting with developers, development time, and paper grades can come to an end.

By extension, if one can modify natural or existing lighting and bring it to some predetermined contrast level, as in the studio, other aspects of image processing will stay constant and precise control over image tonality can be maintained. A system of tone control based on control of lighting contrast works well both with roll-film formats and with color photography wherein variable film development is impractical, but it works only when the subject is nearby and within range of portable lights.

Lighting contrast is not so easily controlled in outdoor scenic photography where the distance between the subject and the background can vary considerably. Nor is it practical to modify lighting contrast in surreptitious modes of photography wherein the photographer must escape detection. Nevertheless, whenever it can be feasibly done, it is the ideal way to control image tonality.

Paper Grade

One can alter print contrast within limits by changing paper grades. Photographic papers are manufactured in various exposure scales, each requiring a negative within a specific density range to bring out the paper's full reflection density. A normal or medium-grade paper, like a number-2 grade, has an exposure scale of about 10. The exposure that creates the darkest tone when printing on such a paper must be 10 times greater than the threshold exposure of the paper. The transmittance ratio of the negative must likewise be about 10:1 or a bit greater to create this range of exposure. If the transmittance ratio of the negative is less than the exposure scale of the paper, print contrast or print-reflection density range will be reduced.

Lens Filters

Lens filters allow refinements over image tonality under specific and limited conditions, such as when certain colors appear in the scene. In black-and-white photography one can change the tonality of a specific object by using a filter of the same color as the object to lighten its tone or of a complementary color to darken it. Polarizing filters can alter contrast in certain scenes by absorbing specular reflections from nonmetallic surfaces. Being neutral in color, polarizing filters work equally well in black-and-white or in color photography. Filters are useful in preventing tonal mergers (discussed later in this chapter and in Chapter 19).

After Treatment

Intensification and reduction are sometimes resorted to as emergency methods of contrast and density correction. Their use is rarely planned since better ways can almost always be found to alter contrast given warning that it will be necessary to do so. Chemical intensification is a process used to increase the density of the silver image and to strengthen contrast in thin negatives. It increases contrast by building density in highlights faster than in shadows. It offers partial correction for underexposure or underdevelopment but only to the extent of strengthening density where density already exists. The process can add no new information to the image.

Reduction lowers the density of a print or negative by bleaching and removing silver particles. It is sometimes used to compensate for overexposure or overdevelopment. This process can actually destroy image information.

Flare

Flare reduces image clarity even when it occurs in small amounts; it is generally out of place in high-resolution work. Flare can be caused by dirty or abraded lenses, defective filters, optical flaws, or uncontrolled light entering the lens and striking the film. Controlled flare is sometimes used deliberately in the darkroom with a technique called flashing to soften print contrast. Beware: flare of any kind reduces the modulation-transfer function and lowers the resolution potential of a system.

Enlarger Light Quality

The various enlarger light sources can alter print contrast significantly. The enlarger light source influences print contrast in proportion to the collimation of its rays. Point-source and condenser enlargers produce greater print contrast than do diffusion and cold-light enlargers. Even if one has lampheads of each kind, changing enlarger light quality from print to print can be awkward. The enlarger is more often treated as a constant in tone-control systems than as a variable; negatives are processed to print well on one kind or another. Yet, when warranted, altering print contrast by changing the enlarger light source is an option that works.

Paper Developer

Print development has little effect on print contrast when prints are developed to finality and for maximum black. When they are not, dilute paper developers can reduce print contrast by the equivalent of about one paper grade. Some photographers never switch print developers except, perhaps, when an isolated negative is too contrasty to be accommodated by the softest graded paper.

Practical Tone Control

Whenever lighting contrast at the subject can be controlled by the photographer, an adequate system of tone control, indeed the preferred system with roll films, is a combination of controlled lighting at the scene and graded paper in the darkroom. Other contrast controls, including development time, are standardized in such a system. Negatives exposed to a nonstandard brightness range may not develop to the density needed to print perfectly on a middle-grade paper, but if all negatives on the roll are correctly exposed and if development time is based on that needed for an average scene, most negatives will print well on another common paper grade.

Modifying lighting contrast from frame to frame in roll films is not only more practical than modifying development, but it may yield greater resolution of detail. Curtailed development effectively reduces the density range of the negative and brings the extremes of lighting under control. At the same time, however, it compresses image tones and reduces separation in middle tonal values. It weakens textural contrast as well. Also, the need to increase film exposure to compensate for curtailed development equates to sacrificing emulsion speed. A better rendition of image features and textures will be achieved by skillful use of light (see Chapter 21).

Chapter 19

High-Resolution Camera Techniques
The High-Resolution Photographer at Work

The secret to making high-quality prints is to begin with high-quality negatives. As obvious as this concept may seem, its importance sometimes goes unappreciated. Some people believe that errors in camera technique can be corrected by fancy darkroom maneuvering. Certain errors can be corrected, but there is no way to recover image detail that is not captured originally in the negative, not even by sophisticated computerized image enhancement.

Poor camera technique is unforgivable because many errors that degrade the camera image are easily controlled. One can control them, however, only by acting before the image is made. Conditions that favor peak resolving power must exist while the shutter is open. This restates an old photographic adage: *capture it on film*. Before tripping the shutter, one must see that the conditions needed to create high-resolution images have been established and that these conditions are sustained until exposure terminates.

Aperture Selection

The principal use of the aperture control is for adjusting exposure. The aperture setting also affects the magnitude of diffraction and the magnitude of certain lens aberrations. Fortunately, one can usually choose from among several combinations of aperture and shutter speed, any of which will yield correct exposure. For example, f/32 at 1/2 second yields the same MCS exposure as f/2 at 1/500th second. Yet these settings are far from interchangeable in regard to their effect on image clarity.

When photographic conditions are ideal, you should consider selecting your lens aperture before selecting your shutter speed. You may initially be inclined to use the aperture that gives peak sharpness, but since you must first identify and act to reduce your limiting spread component, you cannot set the lens to its optimum aperture and leave it at that. Three factors will influence your choice of aperture:

1. Subject depth. The need for an adequately deep zone of sharpness will be the first consideration in selecting an aperture. Your choice should be determined by whether you must have critical sharpness in a single plane of focus, as with a flat subject, or reasonable sharpness over the depth of a three-dimensional subject. Depth of field comes first whenever the subject is three-dimensional.
2. Subject motion. If the subject will move during the exposure, a faster shutter speed may be more important to clarity than peak lens sharpness. Even in small amounts, subject motion can immensely damage image detail and sharpness.
3. Level of illumination. Dim lighting introduces problems that can rule out using the optimum lens aperture. Granted, when the camera is mounted on a solid tripod, you will be able to choose from an almost unlimited range of apertures, one of which will be ideal for your subject, but exposure time becomes increasingly longer as light wanes and ground vibrations eventually become the limiting spread component. Dim lighting can then force the use of a suboptimum aperture, even when the camera is solidly supported.

Optimum
Aperture
Setting

Somewhere near the middle of the aperture scale the spread component from diffraction and from optical aberrations are small enough that their combined spreading is less than it is at the extremes of the aperture scale. At this optimum aperture, lens sharpness is at its best. See Figure 19.1. Although the f-number of the optimum aperture varies from one lens to another, it is often about two or three f-stops from the maximum aperture. It can be found from resolving-power tests or from the modulation-transfer function, and it is very much worthwhile that you know the optimum aperture of every lens you use. See Figure 19.2.

To understand how one might exploit the optimum sharpness of a lens, consider a lens whose optimum aperture falls between f/5.6 and f/8 so that either aperture gives about the same peak sharpness. The better aperture to use for photographing a three-dimensional subject would be f/8. This aperture gives moderate depth of field and in bright lighting allows use of a fast shutter speed. For example, using a slow, ISO 25-speed film at f/8, the shutter speed for a subject

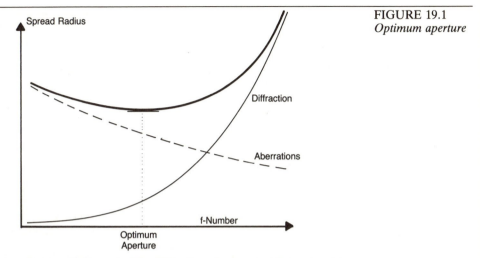

FIGURE 19.1
Optimum aperture

The spread component created by diffraction decreases at large apertures. The spread component created by aberrations decreases at small apertures. At some middle aperture the combined spreading from both is less than it is at either extreme.

in direct sunlight would be 1/125th second, fast enough to control moderate motion blurring. If a faster shutter speed is needed, f/5.6 at 1/250th second is a combination that can be used with no loss in sharpness and with only a slight reduction in depth of field. If the subject is flat and is parallel to the film plane so that extensive depth of field is not needed, you should select f/5.6 from the start, as it allows use of the faster shutter speed with no sacrifice in peak sharpness.

If a subject is deeper than the depth of field given by the optimum aperture, a narrow zone of sharpness will be surrounded by areas that are badly blurred. Although human tolerance for image blurring can be surprisingly high, blurring like this, occurring in the main subject and emphasized by areas of uneven sharpness, is less acceptable. In photographs of full-frame, three-dimensional subjects, it is better that a small aperture be used so that the image is recorded with even sharpness throughout its depth. Although peak resolving power will be sacrificed, consistency of sharpness is often more important to general clarity than impeccable sharpness mixed with massive blurring.

Small Aperture Settings

The disadvantage of using an aperture smaller than optimum is that diffraction negates some of the improved clarity from the extended depth of field. This is why you must not use a small aperture automatically for every photograph. The diffraction-spread component at an aperture of f/22 or f/32 can visibly reduce enlargement latitude. At f/32 diffraction is nearly as great as the commonly accepted maximum size allowed in the 35mm format for the circle of confusion.

Large Aperture Settings

Unless your lens is diffraction limited or is uncommonly well corrected, large apertures are to be avoided or resorted to only in difficult circumstances. Still, their use is justified when subject motion forces the use of a fast shutter speed or when illumination is so low that ground vibrations occur. If the subject happens to be flat or far away, or if the lens is well corrected, images made at large aperture can be quite sharp.

Standard Aperture

A few professional photographers claim to make all their photographs at one aperture setting, usually f/8 or f/11. Admittedly, by using one of these aperture settings all the time, they benefit in several ways: (1) they benefit from the sharpness of the lens at these near-optimum apertures; (2) they learn well the performance of the lens at this aperture; (3) they know the lens' depth-of-field characteristics and can manage depth of field instinctively; (4) they expose their film more consistently as they have memorized the shutter speeds to be used in familiar situations and can quickly adjust their exposure when lighting changes.

Despite its advantages, however, this unorthodox and invariant approach to aperture selection is not at all prudent. The aperture setting should be selected in consideration of the subject and the situation. It is necessary at times to abandon the optimum aperture

FIGURE 19.2
Aperture selection

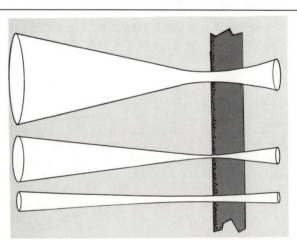

The spread function of a lens is always smallest at an intermediate aperture known as the optimum aperture. It is large at large apertures because of aberrations. It is large at small apertures because of diffraction. At middle apertures, diffraction and aberrations are both reduced and the spread function is smaller than at the extremes.

to obtain suitable depth of field with three-dimensional subjects, to allow the use of a faster shutter speed with moving subjects, or to get reasonable image quality in bad lighting.

Focusing a good lens accurately is more difficult than focusing a poor one. This seeming contradiction comes about because the focal spot formed by a good lens at the point of critical focus is far smaller than can ever be seen without extreme magnification. Even when the image seems perfectly sharp to the unaided eye, sharpness may be degraded somewhat by small focusing errors. The better the lens, the more one must magnify the viewfinder image to focus it with critical sharpness.

Focusing

Although achieving critically sharp focus on a flat subject can be difficult, critical focusing is seldom necessary with ordinary subjects. The requirement in photographs of three-dimensional subjects, for example, is to distribute depth of field, to see that important parts of the subject are within the zone of acceptable sharpness. Of course, when precise focusing does matter, it is important that one understands how to achieve it.

Critical Focusing

Focusing Aids.   Many 35mm single-lens reflex cameras have a 5-power magnifier incorporated into their viewfinders to assist the photographer in focusing accurately. Split-image rangefinder prisms and microprism focusing screens also help. Some camera manufacturers combine the split-image prism and microprisms onto a single focusing screen that may also contain a ground-glass viewing area.

A ground-glass screen is best for previewing depth of field. The split-image rangefinder is easier to use and is the focusing aid most likely to be used accurately. A prism in the split-image rangefinder divides the image into two parts that are aligned when the image is in focus and misaligned when it is not. Split-image focusing is accurate because aligning the shifted image segments depends on the vernier accuracy of vision, which is extremely good. Most observers can detect a misalignment in two line segments that are offset a mere fraction of the point-resolution limit of the eye. A limitation of split-image focusing, however, is that it can be used only at large apertures. The image in a split-image prism blacks out at apertures smaller than f/5.6, rendering the device useless for focusing.

Maximum-Aperture Focusing.   Modern single-lens reflex cameras allow the lens to be focused with the diaphragm fully opened instead of closed down to the taking aperture. This gives a bright viewfinder image for easy focusing. Maximum-aperture focusing can be a trap, however. Complications in lens design occasionally create zonal spherical aberrations in which the magnitude of spherical aberration does not vary smoothly with changes in the aperture setting. Indeed, as the lens is stopped down, such an aberration may actually increase at some point.

Stop-Down Focusing.   For predictable results the lens must be focused at the taking aperture, which may not be easy to do. Even in bright lighting the viewfinder image at small aperture settings is dim. The image will be more easily seen after your eyes adjust to the low light level. It helps to have the camera mounted on a tripod so that you can take your time examining the viewfinder screen. After a while the image will appear brighter. A black focusing cloth that blocks stray light and reflections at the viewfinder will improve the contrast and visibility of the dim image, making it still easier to focus. An auxiliary eyepiece magnifier may or may not help. It will help you to evaluate edge sharpness but will further reduce image brightness.

Parallax Focusing.   In dim lighting, a clear focusing screen marked with a reticle makes for easy focusing because the clear screen provides the brightest possible viewing image. The problem is that the entire image is seen sharply through clear glass. Since out-of-focus points are not blurred as they are on a ground-glass screen, the lens must be focused using a technique called parallax focusing.

To understand how parallax focusing works, think of the aerial image projected by the lens as being a three-dimensional block instead of a flat field. Each object in the scene is represented in this aerial-image block at a distance from the lens determined by the distance of the parent object. An object will be in focus only when the film intersects the object's image in the image block. As a film is moved back and forth through the image block, different planes come into and go out of focus. Actually, 35mm cameras are focused by moving the lens, not the film, so that the image block moves with respect to the film and with respect to the viewfinder. Assuming no alignment problems in the viewfinder system, an object will be rendered sharply at the film plane when its plane in the image block falls exactly on the surface of the viewfinder screen.

This is the key to parallax focusing. The image plane in focus at a particular time is the one that falls on the surface that contains the reticle. Image points that are out of focus fall on a plane in front of or behind the reticle. Out-of-focus points can be easily identified because their position relative to the reticle changes laterally as they are viewed from different points of view. A similar change occurs when one looks through a window. As you move from side to side, the position of objects outside or inside changes with respect to the window frame. A spot on the glass windowpane holds its position with respect to the frame, however. It is the shifting in position of out-of-focus image features that allows one to focus critically. When the image comes precisely into focus, its position remains fixed relative to the reticle when viewed from different points of view. Parallax focusing can be quite accurate, and despite the involved explanation, it is easy to master.

Scale Focusing.   If the distance scale engraved on the lens is accurate or has been specially marked or calibrated, the lens can be focused from the scale by measuring the distance from the camera to the subject with a tape. If carried out with care, this procedure eliminates the visual focusing errors that occur when the subject is too dim to be seen clearly in the viewfinder; it virtually guarantees accurate focusing.

High-Precision Focusing.   Image reductions exceeding 200 diameters are routinely achieved in the reprographics industry. Obtaining such reductions requires superb, precisely focused optics. The full resolving power of the diffraction-limited lenses used to achieve such reductions can be realized only at a large aperture. The extreme sharpness of the lens coupled with the narrow depth of field at the taking aperture make it impossible to focus accurately using conventional techniques. The plane of sharpness cannot be found with ordinary eyepiece magnifiers, and ordinary focusing mechanisms are too crude to position the film precisely at the point of focus.

An alternative method of focusing is used. First the lens is focused roughly using conventional methods. Fine adjustments are then made by moving the subject while examining the image at the film plane with a microscope. Moving the subject instead of the film greatly improves the sensitivity of the system to a slight change in focus. Indeed, a change of one millimeter in the subject's position will change the location of the image plane by about one micron.

A variation to this procedure is to move the camera itself when making the final focusing adjustment. Such a procedure is not easy to use in reprographics applications—the process cameras are too massive—but in 35mm photography it can be a useful method for fine focusing on nearby objects. After the image has been focused in the usual way, move the camera back and forth a fraction of an inch while observing in the viewfinder how image sharpness changes. Then return to and hold the camera at the sharpest position. When you find the plane of sharpness by this method, you can be confident that focus is as accurate as the unaided eye can make it.

Approximate Focusing

You may not have time to focus on a moving subject or during a transient event. To obtain reasonable image sharpness in such situations you must somehow anticipate the action or event and focus in advance.

Prefocusing.   If the subject will appear at a predictable spot or will travel a known course, you can focus in advance on some object near the spot or along the predicted route. For example, to photograph a galloping horse or a racing car, focus on some feature along the track and make the photograph at the instant the subject reaches the selected spot. In sports and in many similarly organized activities, much of the important action can be anticipated. For example, if a runner in baseball advances from third base, key action will be at home plate. In basketball key action is usually near the basket. In nature photography action will often be at watering holes. In each case the lens should be focused on a point near the place where key action is likely to occur. Prefocusing in this manner, while not guaranteed to work, is better than having the camera focused to a random distance. The better you know the subject, the better you can predict its actions, and the better your chances of success with prefocusing.

Zone Focusing.   Zone focusing is useful when you are uncertain beforehand about where the subject will appear or how it will behave. In zone focusing you must allow for a fairly large area or zone in which action might occur, focus somewhere within that zone, and rely on depth of field to bring the image into reasonably sharp focus.

Hyperfocal Focusing.   When a lens is focused at infinity, the near limit of its depth of field, the nearest plane in which focus is acceptable is called the *hyperfocal distance* associated with the lens aperture. If the lens is then refocused to the hyperfocal distance, it acquires a special property: its depth of field will begin at a point halfway between the lens and the hyperfocal distance and will extend to infinity. As it turns out, depth of field is always greatest at a particular aperture when the lens is focused at its hyperfocal distance. The hyperfocal distance of a 50mm lens at an aperture of f/11, for example, is about 4.5 meters. When the lens is focused at this distance, its depth of field will extend from just over 2 meters to infinity—an impressive range.

Image Management

Image management refers to the process of altering the physical or mechanical aspects of an image—its position on the film, its size, or its relationship to the background—to improve its clarity or appeal. Special considerations in image management are control over image size, lens focal-length selection, distribution of depth of field, image placement, and tonal separation.

Full-Frame Composition

For greatest clarity of image features, it is generally best to obtain a large image at the negative stage by filling the film frame with the subject. This is especially important in 35mm and smaller formats where none of the available film area should be wasted. In full-frame photography, all of the negative area is used to reveal the subject and essential elements of the background. Consequently, smaller enlargements are needed and the legibility of the enlarged image is improved. Full-frame composition is facilitated by the through-the-lens viewing systems of single-lens reflex cameras that reveal the subject exactly as it will be recorded on film. Be aware that some 35mm cameras show less in the viewfinder than will be recorded on film. Also be aware that the masks used to mount transparencies may obscure part of the image.

One's choice of lens focal length will be guided by the need for a full-frame image, by the approachability of the subject, and in photographs of three-dimensional subjects by the need for adequate depth of field.

Focal-Length Selection

Medium Focal-Length Lenses.   The strategy for optimizing sharpness in full-frame photography is to position the camera at a distance that fills the frame using the best lens you have. The sharpest lenses available for 35mm cameras typically come in medium focal lengths (35 to 105mm) and may in fact be the prime lens for a particular line of cameras. A medium focal-length lens can be incredibly sharp. (See Plate 7.) But since its optimum aperture usually falls in the range of f/5.6 to f/8, apertures that yield limited depth of field, the peak sharpness of a medium focal-length lens will be realized only with flat or distant subjects.

Telephoto Lenses.   When you are restrained from approaching the subject closely enough to fill the frame using your prime lens, a long focal-length or telephoto lens is called for. A telephoto lens provides a greater magnification ratio and a more spacious working distance than does a normal lens. Dangerous and timid subjects or subjects otherwise unapproachable often create the kind of problem that is best solved by a telephoto lens.

Telephoto lenses should not be used indiscriminately just to get a larger image. These lenses can be difficult to use and they require greater skill in handling than do lenses of shorter focal length. One problem is that, at a given aperture and subject distance, the depth of field of a telephoto lens is narrower than that of a shorter lens. Telephoto lenses also magnify aberrations, blurring, and image defects as much as they magnify the image. In addition, because it is doubly difficult for lens makers to correct the aberrations of fast telephoto lenses, the maximum aperture of a telephoto lens is usually small, about f/8, for example, in a typical 500mm lens. Such small apertures can force the use of slow shutter speeds, which when combined with greater image magnification, increases the danger of motion blurring. Generally, it is best to use a telephoto with the shortest focal length possible, one that fills the frame at the nearest approachable distance.

Wide-Angle Lenses.   At a given aperture and working distance, wide-angle lenses give depth of field unparalleled by that of normal and telephoto lenses. (See Plate 8.) Some of the sharpest photographs you will ever make of three-dimensional subjects will likely be made with wide-angle lenses. Since they do not magnify the image as much as long focal-length lenses do, wide-angle lenses admit less blurring from subject motion. Their wide field of view and greater depth of field are also useful in zone focusing, in uncertain situations, and with unpredictable subjects.

Wide-angle lenses reduce image size. If this means that the negative must be magnified more in the print, the advantages of using the wide-angle lens can be lost. For one thing, blurring from image motion, although less on film than it would have been with a longer lens, will be enlarged back to size in the print. The equivalent blurring will be no different from what it would have been had the longer lens been used from the beginning. Also, much of the clarity gained from depth of field will be lost to graininess when the small negative image is printed at the larger size. It is better to obtain full-frame images at the negative stage so that print enlargements are no greater than necessary.

Depth-of-Field Management

Whenever depth of field is greater than that needed to match the depth of the subject, a smaller aperture or a slower shutter speed than necessary has been used. The small aperture brings more diffraction; the slower shutter speed allows more motion. In either case, image sharpness is likely to decline. Depth of field should therefore be *optimized* rather than maximized. See Figure 19.3.

Depth-of-Field Distribution.   In scenic vistas where the background and foreground are equally important, depth of field must extend to infinity. There is no choice here about the distribution of depth of field. In some photographs, however, only the foreground containing the main subject must be sharp and then only its visible parts. If the far side of the subject will not show in the composition, depth of field that extends from the front of the subject to the center may be adequate. Sometimes the zone of sharpness can be confined to the front third of the subject. By managing the distribution of depth of field in this manner you may find it possible to use an aperture closer to optimum, even with three-dimensional subjects that are photographed full frame.

If available depth of field is shallower than the depth of the subject and you cannot use a smaller aperture to improve it, it is visually more acceptable to distribute sharpness over parts of the subject nearer the camera at the expense of more distant parts. (See Plate 9.) The point of focus should not fall precisely on the front surface of the subject—a common mistake of beginners. To distribute the zone of sharpness squarely about the subject, focus near the subject's center, or even better, slightly forward of center. The ideal point of focus varies with the focal length of the lens and with the camera-to-subject distance.

There is no reason why the composition of a high-resolution image should not be pleasing, but sometimes superior image detail can be obtained by violating formal rules of composition. For example, if an assignment requires that you record a vehicle license plate in poor lighting from a distant position, obtaining adequate resolving power may be challenging. In such situations, where the resolution of detail will be no better than marginal, the critical element, the license plate in this instance, should be placed at the center of the frame where the resolving power of the lens is likely to be greatest.

Image Placement

When subject planes having similar tonal values are adjacent to or overlap one another in a photograph, it becomes difficult to distinguish them. The shape and form of the subject may become confused in the presence of merging tonal values.

Tonal Separation

Obviously tonal mergers must be identified and corrected in advance—before the photograph is made. In black-and-white photography, however, it takes forethought and skill to anticipate such mergers. The subject may contain elements of the same reflectance value but of different color. Although these elements will produce identical shades of gray in a black-and-white print, they will appear different to the eye because of color contrast. There may be no clear indication in the viewfinder of their similarity. Tonal mergers often occur in this way, with no forewarning. Meter readings from a spot-exposure meter can help predict such mergers, but for this technique to work one would have to meter every element in a scene, hardly an efficient process. Special viewing gels that suppress color, such as the Wratten number 90 filter, let you actually see

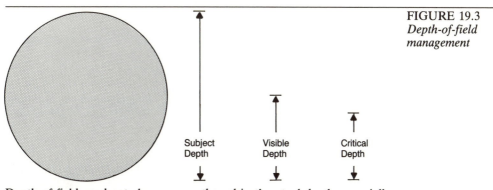

FIGURE 19.3
Depth-of-field management

Subject Depth   Visible Depth   Critical Depth

Depth of field need not always cover the subject's actual depth, especially if the far side of the subject is concealed. If the zone of sharpness extends to the front 40–50 percent of the subject, that is often adequate.

such mergers. This filter mutes chroma and allows you to evaluate the relative brightness of objects independently of their color.

Once you have detected a possible merger you can act to prevent it. A merger can sometimes be avoided by changing the composition. If the subject cannot be moved it may be possible to change the camera angle. Slight adjustments to camera elevation or lateral shifts in its position are often enough to improve tonal separation. If subjects with contrasting colors merge, their tonal relationships can be changed by using a filter similar in color to one subject to lighten it or complementary in color to darken it.

Tonal mergers occur quite often when a subject is dimly illuminated on one side and the background is dark. Mergers of this kind can be eliminated by lighting the subject and its background separately. Background lights are an effective solution to tonal mergers in studio photography. Lighting the subject from behind is another alternative, one that is sometimes as easy to set up in the field as in the studio.

Chapter 20

Controlling Image Displacement
Reducing Blurring from Camera and Subject Motion

Motion blurring can conceivably be reduced below the resolution limit of ordinary camera systems, but doing so can be tricky. Obtaining total and absolute immobility in a camera setup is challenging and may be impossible. To solve the problem of ground vibrations in stellar photography, for example, one astronomer recommends that telescopic cameras be mounted on wooden piers sunk deeply into the ground. Metal or concrete piers are less useful as they transmit vibrations, such as from highways or ground tremors, more readily than does wood. Industry offers another example. To control ground vibrations in micrographics and integrated circuit applications, cameras and easels are combined into a single massive structure. If ground vibrations occur, the structure moves as a unit. In this way a high degree of stability between camera and subject is obtained where absolute immobility cannot be achieved.

Since camera and subject motion are the factors most likely to limit the resolving power a beginning photographer can hope to get, much will be gained by successfully controlling these irritants. Many novices do not understand the importance of tripods and fast shutter speeds in controlling motion blurring; others overestimate their effectiveness. (See Plate 18.) The penalty for such naiveté is a self-imposed resolution limit. That is, the detail and sharpness achieved by an unwary photographer, more likely than not, will be limited by some kind of motion-spread component. Often the motion is transmitted to the camera when it is hand held, but even when the camera is mounted to a tripod, it can be set into motion by external disturbances or by internal vibrations.

Photograph-ing Moving Subjects

Subject motion creates challenging problems for the high-resolution photographer. If a fast shutter speed is to be used, the photographer may not be able to use the aperture needed for adequate depth of field. If a faster film is to be used to make the faster shutter speed possible, grain and turbidity will reduce clarity. Despite the second-ary problems caused by fast shutters and fast films, let there be no doubt that either alternative yields better image quality than will be possible if motion blurring obliterates image detail. The best remedy is the one that minimizes cumulative degradation. Keep in mind that small amounts of motion blurring can degrade an image rapidly.

Before resorting to alternatives that reduce clarity, consider those that do not. For example, you may be able to reduce blurring directly at the film plane, or you may be able to use an indirect approach, such as increasing the illumination on the subject, panning the camera, or waiting for the peak of action.

Direct Reduction of Image Blurring

Your success in directly controlling the magnitude of blurring from subject motion will depend on your ability to control the variables in the equation derived in Chapter 15:

$$d = Mtm(\cos a)$$

where d represents the displacement blur component, M is image magnification, t is exposure time, m is the speed of subject, and a is the angle of the motion with respect to the film plane.

Some of these variables may be beyond control. For example, you may not be able to control the speed and direction of subject motion. Still, since the angular speed of the subject is what matters, there may be another way. If you know in advance how the subject will move, you can select a vantage point that records it moving toward or away from the camera instead of parallel to the film plane. From such vantage points blurring will be reduced noticeably.

The remaining factors, magnification and shutter speed, offer the photographer the best opportunity directly to control subject motion blurring. Magnification varies with lens focal length and camera-to-subject distance, both of which can be selected by the photographer. Long focal-length lenses and near subject distances increase image size and increase the magnitude of motion blurring. Conversely, wide-angle lenses and distant subject positions reduce image size and reduce blurring. But as stated in the last chapter, it does not help to reduce magnification at the camera stage if the image is to be enlarged substantially in the print. Blurring is increased by mag-nification without regard to whether the larger image is obtained in the camera or in the darkroom.

The blurring caused by camera or subject motion is probably the most insidious and beguiling of image-degrading spread factors. It occurs far more frequently than it should, no doubt due to the widespread and erroneous belief that fast shutter speeds "stop" image motion. When an image moves during exposure, the reality of its motion cannot be obscured at any shutter speed. Fast shutters limit the distance the image can move in the brief instant the shutter is open and thereby limit the blurring that registers on film. As long as the object moves, however, the image moves. Fast shutters lessen blurring and ultimately reduce it below the resolution potential of the system. Even then, blurring has not been eliminated.

Finding a "safe" shutter speed can be tricky. As seen in Part I, spread components combine to degrade image quality cumulatively. If the spread function of the system before being acted on by a motion component is just marginally below threshold size, even a small motion component can bring the total above threshold. A good guideline is to always use the fastest shutter speed that you can without violating principles of sound exposure and aperture selection. A more important rule is that the largest source of degradation in an image should be reduced first. When motion blurring becomes the spread component that limits resolving power, reducing such blurring must take priority.

Special Shutters.   Cameras with built-in shutter speeds of 1/1000th of a second (one millisecond) are fairly common. Indeed, built-in shutter speeds of 1/8000th of a second (125 microseconds) are available in a few new models. If shutter speeds faster than these are needed, specialized shutters must be used. As indicated in Table 20.1, several kinds of high-speed shutters are available, some of which provide exposure times in the microsecond range and shorter.

Fast shutters simplify matters only when the subject is well illuminated or when supplemental lighting can be used. They may also require the use of fast films and large apertures as discussed earlier. Ultra-fast shutters thus complicate the task of making high-resolution images by requiring bulky lights, low-resolution films, suboptimal apertures, or a combination of these. It is often simpler to dispense with special shutters and use high-speed electronic flash units instead.

Of the indirect ways to combat blurring from subject motion, the best is bright lighting (see Chapter 21). The simple advantage of good illumination is that it allows for both a short exposure time

Shutter-Speed Selection

Increasing the Light Level

Shutter type	Speeds available		
Magneto-optical	12–20	microseconds	
Electro-optical	7	microseconds to	
	7	nanoseconds	
Image tubes	0.005	microseconds	
Image-converter tubes	0.02	microseconds	
Oconoscopes	2	microseconds	

TABLE 20.1
High-speed shutters

and a more ideal aperture setting. If you are already using supplementary lights, you can place the lights closer to the subject, use units of greater power, or use additional units.

Panning   It is possible to compensate for certain kinds of subject motion by panning the camera so that the film moves at the same angular speed as the subject. See Figure 20.1. However, panning works only when subject motion is smooth and predictable. To use the technique one must track the subject in the viewfinder and keep the image in the same position on film throughout the exposure. When the camera is panned correctly, the subject will be rendered sharply against a blurred background.

Panning is a skill that takes practice to perfect. It can be quite difficult to track a fast-moving object accurately and smoothly. One may not, for example, be able to keep a low-flying airplane or a speeding race car in the viewfinder at all, much less keep them in the same position relative to the film. With single-lens reflex cameras, it is virtually impossible to track such subjects without an external viewfinder or a sportsfinder, because the reflex-viewing system blacks out during exposure after the viewing mirror swings out of the image-forming path.

To pan smoothly, start panning before releasing the shutter, then continue to pan until you are certain that the shutter has closed. See Figure 20.2. Avoid jerky starts and stops, as this erratic motion may carry over into the actual exposure. Also be aware that objects moving parallel to the film plane will show an apparent increase in linear speed or an actual increase in angular speed as they approach the camera. To track the object accurately throughout its path, one must pan faster as the object approaches the camera position, panning more slowly again as it recedes.

FIGURE 20.1
Panning-speed
differential

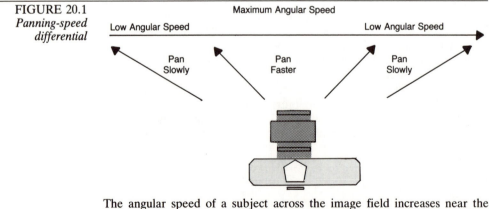

The angular speed of a subject across the image field increases near the camera position. The panning speed must vary if the subject is to be tracked accurately.

Mounting the camera on a steady tripod helps both in framing the subject and in maintaining a smooth panning motion. Pan heads allow the camera to turn freely about the vertical axis and are available on most tripods. Fluid-damped pan heads allow a very smooth panning action because they resist too rapid a change in panning speed and prevent the operator from jerking the camera.

Moving objects that reverse direction at predictable points, such as when jumping, swinging, or bouncing, can be rendered with less blurring if photographed at the height of the motion. (See Plate 10.) Such objects always slow down somewhat as they change direction at the peak. This is where they should be photographed. If the motion is exactly vertical, for example, the subject will come to a momentary stop before its direction changes. To use this technique, you must anticipate the peak of action and allow for the brief lag in your reaction time between the moment you decide to act and the time your body actually responds. You must thus begin to release the shutter a fraction of a second before the peak of action occurs.

Peak of Action

The photographer gains such incomparable advantages from the correct use of a tripod that its use should be considered mandatory unless prevented by compelling circumstances. Some of these advantages are listed below.

Tripods and Other Camera Supports

Advantages of Tripods

Aperture Control.   To get maximum clarity of detail, you must select an aperture suited to the depth of the subject. A tripod, by giving stability to the camera at slow shutter speeds, extends the range of usable shutter speeds and apertures. Even in dim lighting a tripod may let you select an aperture for peak sharpness or for extended depth of field as conditions warrant.

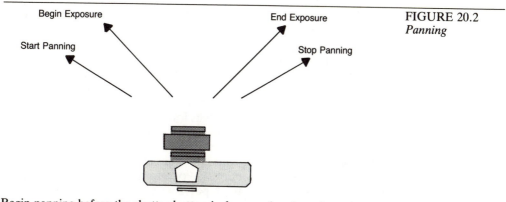

Begin Exposure

End Exposure

Start Panning

Stop Panning

FIGURE 20.2
Panning

Begin panning before the shutter button is depressed and continue panning until the shutter closes. Developing a good technique takes practice.

Depth-of-Field Management.   A tripod lets you control the distribution of depth of field more precisely. To see in the viewfinder how the zone of sharpness is distributed over the subject, you must view the image with the lens aperture stopped down. This is easier when the camera is held at one position long enough for your eyes to adjust to the dim image.

Precise Focusing.   Tripods make critical focusing easier, especially in close-up shots where the zone of sharpness shifts significantly with minor changes in camera position.

Image Composition.   The habitual, careful, and deliberate use of a tripod will condition you to compose with greater care. The small effort it takes to relocate the tripod will motivate you to think more carefully before setting up. Careful composition, in addition to improving the visual appeal of the image, will result in less cropping of prints, smaller enlargements, and in turn, finer grain and better print tonality.

Minimum Image Blurring.   Tripods, of course, help you to eliminate certain kinds of image motion, especially the severe kinds associated with hand-held cameras. They can also help you control internal camera vibrations caused by the focal-plane shutter and viewing mirror of single-lens reflex cameras.

Selecting a Tripod

Even if you have a small, low-cost, low-quality camera, using a sturdy, rigid tripod is a fast and certain way to improve the sharpness of your images. Lightweight tripods, even when used with lightweight cameras, will do less to improve image quality than you may suppose. Though it is logical that large cameras need strong tripods, it does not follow that small cameras can get by with weak ones.

From promotional literature it appears that tripods intended for the amateur market are designed for portability, compactness, and lightness. Unfortunately, none of these properties add in the least to the performance of a tripod.

For a tripod to protect against image displacement, it must be rigid, strong, and stable. A tripod is *stable* when it resists a change in position: it resists overturning and swaying in reaction to wind and ground vibrations. Stability is related both to the mass of the structure and to its center of gravity. Stability improves as mass increases, as the center of gravity is lowered, and as the base formed by the tripod's legs widens. The stability of the wooden tripods used by cinematographers cannot be beaten. Of course, they are quite heavy compared with the tubular aluminum tripods used by still photographers, not to mention bulky and awkward to carry. Their weight obviously reduces their portability, but their stability is unparalleled.

A properly engineered aluminum tripod can have less weight and bulk than a wooden tripod and be as strong and rigid. It may not be as stable, but a tripod made of heavy-gauge, heat-treated, tubular aluminum will aid in producing immaculate images if its weight has not been reduced at the expense of its sturdiness.

To be *rigid,* a tripod must have no looseness in its connecting parts or weakness in its supporting members. The structure should not bend or bow under moderate stress. The legs should not flex or spring. Once adjustments are set they should stay set. Rigidity is related to the strength of the materials used and is related to the design of leg locks. Lever-operated leg locks should be examined carefully to see whether they hold strongly. Friction locks become stronger with size. The greater the locking frictional area the less force needed to maintain them in their locked position and the more rigid the locked structure becomes.

The *strength* of a tripod depends partly on the geometric shape and thickness of the walls of its leg sections. Tubular construction is stronger than channel construction at a given wall thickness. Tripods made of thin sheet metal stamped into channel shapes often perform poorly. They flex easily when jarred and they may resonate in reaction to slight vibrations.

Resonance is the sonic phenomenon that causes a bell to ring. If a tripod resonates, a sharp impact will cause it to vibrate as a chime would. A resonant structure, rather than resisting and damping the vibrations of the camera, will amplify and perpetuate them. Lightweight, resonant tripods offer little protection against microvibrations. Indeed, using a flimsy tripod can be worse than using none at all.

A tripod purchased for high-resolution work should be tested for its rigidity and strength before being accepted. It should resist giving way to moderate downward pressure when the legs are extended and locked. A spongy feel is a good indication that the tripod will be of marginal service. Such a tripod may be adequate under calm conditions when used with quiet, vibration-free rangefinder cameras, but useless when used with single-lens reflex cameras, providing scant improvement over hand holding.

Other Camera Supports

Camera clamps offer a lightweight, highly portable method of supporting a camera. They give very solid support provided they are strong and can be rigidly locked, and provided there is a strong, rigid object nearby to which they can be clamped.

Monopods, chainpods, and gunpods aid in hand holding to some degree. If they do nothing else they encourage the photographer to think about steadiness. For the most part, such devices offer little stability and can at best restrict motion in one direction. Otherwise, they allow complete freedom of motion. They do little to counteract the motion of the pulse and other body movements.

Monopods must be as strong and rigid as tripods to be of any use. A monopod that flexes is useless because it allows the camera to move in all directions, unrestricted. Rigid monopods can be beneficial in situations where the camera must be panned to follow fast-moving action or in confined areas where setting up a tripod would be awkward.

A chainpod is a strong cord or chain connected on one end to the camera; the other end hangs loose. To use a chainpod the photographer stands on the loose end and pulls the chain taut. The premise is that camera motion will be reduced as long as tension is kept on the chain. In practice the camera moves laterally and rotates as freely as a hand-held camera.

Gunpods are useful in action photography. They offer mobility and allow somewhat better control over camera motion than hand holding. The improved stability offered by a gunpod is particularly useful with a telephoto lens or when the camera must be panned. Gunpods cannot replace a tripod when maximum stability is needed, however.

Tripod Techniques

A tripod can be used as a free-standing camera support or as an aid to hand holding a camera. When used as a free-standing support the camera is mounted securely to the tripod head and a cable release is used to trip the shutter. When used as a hand-holding aid, the camera is actually held or touched in some way while being supported on the tripod. The goal in either case is to prevent the camera from moving during the exposure.

Free-Standing Tripods

A free-standing tripod prevents camera movement quite effectively when used with rangefinder cameras or with cameras having leaf shutters. Used with single-lens reflex cameras, a free-standing tripod is not entirely steady. As soon as the shutter of a single-lens reflex camera is released, the camera vibrates from the action of its internal mechanisms. Lightweight cameras used with lightweight tripods are especially unstable. Not only is there the danger of internal motion, but the gentlest breeze or slightest disturbance can set them in motion. If such an unstable structure must be used, it should not be touched while an exposure is in progress.

Cable Release.   A cable release is a useful accessory, but the wrong kind can do as much harm as good. Short, stiff cable releases transmit motion; long, flexible cable releases suppress it. A cable release will transmit vibrations if it is rigid enough to stand erectly. Touching such a cable release will be no safer than touching the shutter-release button directly. A cable release with a stiff wire center must be about 18 inches long and flexible enough to form a loop if it is to suppress vibrations effectively. Pneumatic cable releases made of soft, flexible rubber air hose are much safer than the wire kind.

Indeed, they completely insulate the camera from the motion of the photographer.

Mirror Lock-Up. Tripods and cable releases do not diminish the vibrations caused by the internal mechanisms of single-lens reflex cameras. Still the vibrations set off by the motion of the mirror can be eliminated if your camera allows you to lock the mirror in its horizontal position before the exposure is made. A disadvantage of this procedure is that the image blacks out while the mirror is locked up and renders the reflex viewing system useless until the mirror is released. Before you lock the mirror the image must be composed, framed, and focused exactly as you want it, or an external viewfinder must be used.

Reducing Shutter Vibration. The focal-plane shutter commonly used in single-lens reflex cameras is the second source of internal vibrations. A free-standing tripod, except for the damping mass it adds to the camera, does little to reduce this kind of vibration. The damage done by shutter vibrations will depend on the exact shutter speed used. In the shutter-speed range faster than about 1/125th or 1/250th second, shutter vibrations have no effect. At such speeds, both blades of a typical shutter will have fully closed and exposure will have terminated before shutter vibrations begin.

During time exposures longer than one or two seconds, shutter vibrations can be prevented by by-passing the mechanical shutter, that is, by manually timing the exposure at the T or B shutter settings using a lens cap to start and stop exposure. After removing the lens cap, continue to block the exposing light for a few seconds while the camera settles down, then make the exposure.

Shutter vibrations are thus most troublesome in the range of shutter speeds between 1/2 second to about 1/60th second. Shutter vibrations can be suppressed in this range of shutter speeds by loading or stressing the tripod.

Loading and Stressing. The inertia and stability of a tripod are improved by adding weight to it; the extra weight improves stability by lowering the center of gravity. If the added weight is centered on the tripod, the center of gravity is further lowered and stability is even better.

Adding mass to the structure also increases its effectiveness in damping vibrations. Sandbags, beanbags, bricks, water jugs, and other deadweight all improve damping and stability. An old trick used by professionals is to suspend a camera bag from the center post. Another solution is to carry a length of cord or chain along and suspend some heavy found object from the tripod. The tripod must, of course, be strong enough to support the added weight and rigid enough to support it without getting springy. (See Plate 12.)

Sandbags or bags filled with lead shot, used in conjunction with a sturdy tripod, aid in creating images of immaculate sharpness. Sandbags improve the stability of the tripod structure while absorbing much of the vibration caused by camera mechanisms. The bags may be placed over or under the camera or may be draped around a long focal-length telephoto lens. Sandbags are particularly useful in field photography and can substitute for a tripod when a tripod cannot be used for one reason or another. When sandbags are used without a tripod, the bags may be supported on a tree branch, fence post, or any available sturdy object. Greatest stability is achieved by using two bags, one below the camera molded to fit it snugly, the other above it to improve stability and damping.

Aid to Hand Holding

If your tripod can support your weight without flexing, you may use your own body weight to load and stress the tripod. A workable procedure is simply to hold the tripod-mounted camera while exerting a slight downward pressure with your hands. This technique will not work with a weak tripod. If the tripod flexes it will oppose the downward pressure with a spring-like tension. The technique will then yield no better results than hand holding. If the tripod flexes even slightly, touching the camera may lead to noticeable blurring. Even with rigid tripods it is risky to touch the tripod or camera when using shutter speeds slower than about 1/4th to 1/8th second.

Hand Holding a Camera

Despite their resolve, photographers do not always use tripods. There are circumstances when, for various legitimate reasons, tripods are dispensed with and image quality is knowingly sacrificed. This often happens in fast-breaking situations where tripods are totally impractical. If no alternative means of supporting the camera can be found, the camera is simply hand held.

Gripping the Camera

The obvious kinds of camera shake can be minimized by adopting a good hand-holding technique. The best grip is functional, immobilizes the camera, and allows easy access to camera controls. You should work out the details of a steady and convenient grip, one that suits your shooting style and works well with your camera. Your camera's instruction manual may offer good, solid tips about methods for hand holding your camera.

Cameras with eye-level viewfinders, which includes most 35mm cameras, should be held to the eye so that the camera or the hand that cradles it makes solid contact with the cheek and forehead. Steady the camera with both hands using a firm, not tense, grip. Keep both arms close to your body and motionless. Such a position may seem cramped and unnatural at first, but it will become more comfortable in time. In making the exposure, squeeze the shutter-release button gently and smoothly. Avoid roughness or jerkiness that might cause unwanted motion at the instant of exposure.

You can improve your stability and that of the camera by bracing yourself against a wall, tree, or other solid object if you happen to be near one. The greatest improvement will be obtained when the camera is placed against the object directly. Kneeling and prone positions may be used to advantage also. Kneeling on one knee, brace the elbow of the arm that supports the camera against the elevated knee. In this way you will provide a continuous line of support from the ground to the camera through the bones of your arm and leg. The prone position is steadier than a kneeling or standing position. In a prone position your body and two elbows make solid contact with the ground giving fairly stable support.

Table 20.2 illustrates how quickly quality can be lost when a camera is hand-held. This table shows the theoretical reduction in the resolving power of two systems when a motion of 0.2mm (200 microns), such as can be caused by pulse displacement, occurs in 1/10th second. Resolving power when no motion occurs is assumed to be a maximum of 40 lines per millimeter in the first system and 100 lines per millimeter in the second. Table 20.2 helps clarify a few misconceptions about the minimum safe shutter speed, about the performance of a camera when it is hand held, and about the effectiveness of the shutter in eliminating the degradation caused by image motion.

Image Quality in Hand-Held Photography

Minimum Safe Shutter Speed.   At a shutter speed of 1/250th second, the resolving power of the 40-line system in Table 20.2 is nearly as good as it can become, being only 5 percent below its maximum. At faster shutter speeds, image quality does not differ greatly from the system's peak quality. Many photographers will consider 1/250th second a safe speed to use for hand holding the lower-quality system. This shutter speed, as you might guess, should also produce close to peak quality when used with systems resolving less than 40 lines per millimeter.

The same shutter speed is certainly inadequate, however, with systems of higher quality. The resolving power of the 100-line system at 1/250th second is 22 percent below its maximum. The loss of quality in this system continues to be significant at shutter speeds traditionally considered safe for hand holding. Clearly the shutter speed needed to render image motion imperceptible depends on the potential resolving power of the system as much as on the magnitude of the blurring. Systems of high quality are more sensitive than systems of low quality to the effect of an identical amount of motion blurring.

1/10	1/15	1/30	1/60	1/125	1/250	1/500	1/1000	Max RP
5	7	15	25	34	38	39.5	39.9	40
5	7	16	30	53	78	93	98	100

TABLE 20.2
Resolving power of hand-held cameras at various shutter speeds

Equivalent Performance.   One can also see from Table 20.2 that, at slow shutter speeds, there is essentially no difference between the images produced by a high-quality hand-held system and those produced by a low-quality hand-held system. At shutter speeds below 1/30th second, the quality achieved with the 100-line system is indistinguishable from that achieved with the 40-line system. A photographer loses the advantage of having a high-resolution system when hand holding the camera at a slow shutter speed.

Degradation at All Speeds.   Finally, Table 20.2 confirms that resolving power is sacrificed at all shutter speeds when high-quality systems are hand held. It is pointless to ask whether a hand-held image will be degraded—the question is by how much. How much legibility will one sacrifice to achieve freedom and mobility?

Chapter 21

Photographic Lighting
Using Light to Resolve Fine Detail

Chapter 9 showed how the quality, direction, intensity, and color of light can enhance or degrade image clarity. This chapter shows how light is manipulated to good advantage, how image contrast and gradation can be controlled at the camera stage, how natural lighting can be simulated using artificial lights, and how the resolution of textural detail can be improved by a careful choice and balance in lighting quality, direction, and contrast.

Lighting Contrast

Lighting contrast is used for tone control as well as for enhancing image clarity. In both applications, powerful, portable electronic flash units and lightweight reflectors offer photographers tremendous flexibility and control.

A first step in controlling lighting contrast is to evaluate ambient lighting as to whether existing contrast is satisfactory. If it is not, you must decide how and by how much it must be changed. To measure lighting contrast with an incident-light exposure meter, take direct readings in the brightest light and in shade and count the number of f-stops by which the two readings differ. A difference of one f-stop indicates a 2:1 contrast ratio, two f-stops, 4:1, three f-stops, 8:1, and so forth. If you use a built-in camera meter or a reflected-light meter, take your light readings from a middle-gray reflectance. Most well-stocked camera stores carry 18 percent gray cards in various sizes that are excellent for taking reflected-light exposure readings. Take one reading with the card in direct lighting and the other with the card in the shade.

You must often take outdoor lighting contrast as you find it, but if the subject is nearby, contrast can be increased or decreased in several ways. Excessive lighting contrast can be reduced by putting more light in shaded areas using reflectors or supplemental lights. In flat lighting, as on an overcast day, lighting contrast can be increased by using a supplemental flash unit as a main light to build highlight luminance. This is *additive lighting*. Alternatively, part of the light falling on the subject on an overcast day can be blocked using a black, non-reflecting board to create shading. This is *subtractive lighting*. How subtractive lighting works on an overcast day may not be apparent until you realize that the overcast sky is the light source. The purpose of the nonreflecting board then becomes to prevent skylight from reaching the part of the subject you want shaded.

In studios or wherever a scene can be lighted artificially, you can establish any lighting contrast you want. Table 21.1 will help you to establish specific lighting and contrast ratios when you use two lighting units of equal power. To obtain the lighting ratio in column I or the contrast ratio in column II, simply adjust the distance between the lights and the subject so that the secondary light is farther than the primary light by the distance ratio in column III. If the main light is 5 feet from the subject and you want a 3:1 contrast ratio, place the secondary light 7 feet from the subject (1.4 times 5 feet).

Principles of Natural Lighting

The most enduring principles of photographic lighting are those whose application yields a natural or credible lighting effect. By "natural" is meant the kind of lighting observed in nature, specifically in daylight. By "credible" is meant lighting which is believable or likely. Three of the most widely accepted of these lighting principles are discussed below.

Overhead Source

If natural lighting is to be simulated, the principle source must strike the subject from above at a moderate angle. One usually sees by the light of the sun, which always comes from above. In many temperate areas of the world the midday sun hovers overhead at an angle near 30 to 60 degrees. Lighting from such angles, being of the kind people are most accustomed to seeing, produces the most

TABLE 21.1
Lighting ratio and lighting contrast

I. Lighting ratio	II. Contrast ratio	III. Distance ratio
1:1	2:1	1.0
2:1	3:1	1.4
3:1	4:1	1.7
4:1	5:1	2.0

Note: To obtain the lighting ratio in column I or the contrast ratio in column II, adjust the lamp-to-subject distance so that the secondary light is farther than the primary light by the distance ratio in column III.

natural and comforting visual effect. Since these angles also improve the rendition of textural detail, this principle is consistent with the goals of the high-resolution photographer.

Shadow Detail

A second lighting principle maintains that visible details should appear throughout a photograph, including shadow areas. Even in contrasty daylight, the eye can always discern details in the scene, both in sunlit and shaded areas. One therefore expects to see details in both sunlit and shaded areas in photographs. For this reason, a photograph containing large masses of deep shadows, devoid of detail, can be unsettling. To guarantee that a photograph has adequate shadow detail, important shaded areas must often be illuminated separately using fill lighting or reflectors.

Single Source

A third principle maintains that a photograph should appear to have been illuminated by one light source and that only one set of shadows should be visible in the picture area. When a lighting scheme uses multiple light sources, secondary lighting units should be placed so that confusing or contradictory shadows cannot be seen from the camera position. Visible shadows should support the idea that a single dominant light source was used.

While lighting based on this principle may seem natural, such lighting offers no special advantage in high-resolution photography. The presence of multiple shadows does not in itself lower image detail or sharpness unless the shadows are exaggerated and fall in important image areas. Indeed, this principle encourages the use of axial fill lighting, which eliminates the secondary shadows commonly created in multiple lighting schemes, but which weakens textural contrast.

Lighting for Clarity and Detail

Lighting schemes used by professional photographers are not always designed to bring out detail. Some photographers seek to glamorize the subject and enhance its emotional appeal by whatever means necessary. Many of the lighting schemes used in portraiture and in certain kinds of commercial photography deliberately conceal weak subject features by eliminating detail. They may or may not suit the needs of the high-resolution photographer.

The kind of lighting needed to enhance clarity in a photograph depends partly on the structural characteristics of the subject: whether its shape is flat or contoured, its texture fine or coarse, or its sheen matte or glossy. As a guide to determining a basic lighting scheme, the shapes of three-dimensional objects are categorized as either cubical or spherical.

By *cubical* is meant objects that are roughly box-like, having surfaces more flat than curved. The usual practice in lighting cubical objects is to vary the illumination on each plane so that each is isolated tonally from the others. The objective is to achieve variation in light and shade that defines and separates the planes of the subject. Buildings, machines, and many manmade objects are often more cubical than spherical and thereby benefit from cubical lighting.

By *spherical* is meant objects that are rounded or have contoured surfaces. Many such objects are reproduced best in angular lighting, which creates modeling shadows that reveal the subject's shape in relief. Natural subjects—people, animals, geological formations, and so forth—are often more contoured than cubical and benefit from angular lighting.

The surface characteristics of the subject also play a role in determining one's lighting scheme. Surfaces may be textured or plain, glossy or matte. A golf ball, for example, is both glossy and coarsely textured with an embossure. Writing paper is matte and so finely textured as to be plain.

Lighting for Texture

One can exploit the properties of light to enhance textural detail in a photograph by applying three principles: use *directional lighting* to create textural shadows with sharp edges, use *angular lighting* to create textural shadows of adequate size, and maintain acceptable lighting *contrast*.

Directional Light.   Anyone who intends to use directional lighting in professional applications should consider using incandescent spotlights, even though they can be inconvenient. Their overriding advantage is that they allow the photographer to observe the exact lighting effect achieved before the exposure is made. Modeling lights on electronic flash units allow the general lighting effect to be observed, but because a separate light source is used for modeling, there may be slight differences in the illumination pattern of the flash tube.

Other sources of directional light are also available for those who cannot justify the purchase of professional spots. Sunlight is a free, convenient, and often neglected source of directional light. Movie lights, miniature spots, and even slide projectors are other readily available sources. Also, it is easy to simulate point-source lights. Incandescent bulbs with small filaments and almost any lighting unit designed to be used at great camera-to-subject distances operate like point sources. Many small electronic flash units, when placed at least a yard or two from the subject, are point sources for practical purposes and produce shadows with very sharp outlines. Indeed, if a small electronic flash unit has an efficient reflector, its shadows may be sharp at distances as close as one or two feet.

Controlling Textural Contrast.   A lighting scheme that enhances textural contrast without degrading general contrast is quite useful in high-resolution photography. When textural contrast is increased using one of the methods discussed in Chapter 18, however, general contrast is increased equally. The simplest way to control them separately is by skillful lighting.

In effect, a dichotomy of contrast is achieved by illuminating important textural surfaces independently using an angular, directional main light and moderate fill lighting. Contrast between the main subject and the background can be fixed using a lighting ratio of 3:1 to obtain good shadow detail, while local contrast is raised using a lighting ratio of 5:1 or higher for improved visibility of detail in the microstructure of textured surfaces.

The technique is well worth the effort it takes to set it up. It is used frequently in commercial photography where enhanced textural detail adds considerably to product appeal and creates a more intimate view of the featured product. Sometimes separate lighting schemes are used, one to reveal surface texture, another to establish background lighting and general contrast. The light dedicated to the textured surfaces of the product is typically a glancing, angular light. Lighting on the background must be directed so that it brings general contrast to the desired level but does not dilute textural contrast at the center of interest.

Angular lighting is used to clarify the three-dimensional shape of spherical or contoured objects. Angular lighting creates variations in shading across a surface that define its topology. It is sometimes called modeling light for the way it clarifies form, heightens dimensionality, and enhances the illusion of depth.

Lighting for Form and Contour

Backlighting.   Backlighting achieves modeling like that of angular lighting, achieves some of the textural enhancements of directional lighting, and adds brilliance to the subject's outline, setting it off more clearly from the background. (See Plate 11.) The foreground shadow created by backlighting further clarifies the subject's shape and form by repeating it. Backlighting in this context does not mean that the light is placed axially behind the subject where it is hidden, or that it is placed within the lens' field of view where it would create flare. Indeed, the useful effects of backlighting can be achieved by placing the main light just slightly to the rear of the subject plane.

Glare Lighting.   Controlled specular lighting or glare lighting is a special application of backlighting that can be useful for illuminating glossy or shiny surfaces. In industrial photography, for example, the use of frontal lighting to illuminate dark background machines can be frustrating. Light that strikes the metallic surfaces will glance off, away from the camera, reflecting almost fully as specular light. Such surfaces are better illuminated from behind so that specular glare is directed toward the camera. Indeed, just a small amount of glare is enough to bring brilliance to such surfaces.

Lighting Glossy Surfaces   Shiny mirror-like surfaces, particularly those that are smooth, non-textured, and unembossed, require special lighting. A mirror does not disperse light as a matte surface does, but it reflects the image of its surroundings, sometimes merging with the surroundings in tone. A mirror-like object surrounded by dark, nonreflective walls will reflect an image of the light source and will otherwise reflect the dark walls. Its surface tone will not differ greatly from that of the background it reflects, and the object may not be easily seen.

Tent Lighting.   Glossy objects can be illuminated over their full surface area and can thus be made visible by means of diffuse tent lighting. Tent lighting is achieved by surrounding the subject with large reflecting or translucent surfaces that enclose the subject like a tent. Broad diffuse lights are then directed onto the tent-like structure so that the subject is illuminated indirectly by the tent. The subject becomes visible in effect by reflecting the image of the tent.

Axial Lighting   Axial lighting is useful in photographing cracks, fractures, caverns, and deeply fissured subjects. It is frequently used in medical and dental photography to explore recessed organs and body cavities. Axial lighting tends to produce intense, saturated colors. It is excellent for photographing subjects whose contrasting colors define and separate the details of the subject.

Precise axial lighting has some specialized uses. For example, it simplifies the task of making accurate photographs of latent fingerprints on a shiny, reflective surface. Tent lighting does not work because, when such a subject is aligned parallel to the film plane for an undistorted view, it reflects an image of the lens instead of the tent. Precise axial lighting creates pure specular reflections on the reflective surface while the powdered fingerprint interrupts the reflections to provide a visible and accurate fingerprint image.

Printed circuit boards are often rendered more clearly under axial lighting. These boards typically have intricate patterns etched onto their surfaces and have other components mounted above the surface. Angular lighting causes shadows to be cast by the raised components, obscuring and confusing the circuit patterns below. Precise axial lighting eliminates the shadows and reveals the structure of both raised and surface features.

Camera-mounted flash units, which are notorious for their peripheral shadows, give near-axial lighting. Such lighting is sometimes used to good advantage in fast-paced color photography, but even then it works best when the peripheral shadows are not obvious. Mounting the flash precisely above the lens places these shadows below the subject where they are less conspicuous. If you must use a camera-mounted flash unit, place the lamp high above the lens, centered on the lens axis.

Bright sunlight, if not the best light source for high-resolution photography, ranks high on the list. Its principal advantages, of course, are its directionality and intensity. Although natural lighting does have certain disadvantages, you would do well to work in bright sunlight whenever you can.

The obvious problem with direct sunlight is the contrast it gives. The same options used in the studio to control lighting contrast—fill lights and reflectors—are used in daylight photography when practical, but these options are of no use when the subject is beyond the effective illumination range of the fill lights.

Another problem with natural sunlight is that, except for small and nearby subjects, one cannot directly alter its quality or direction. Nevertheless, one may still be able to arrive at a specific contrast, direction, or quality in natural daylight—with planning and patience.

In many geographic areas it is possible to obtain natural lighting of several kinds, directional or diffuse, angular or axial, provided one waits for the right conditions to occur. This generally works best when one can predict the time of day, the season of the year, and the specific atmospheric conditions that provide the kind of lighting desired.

Occasionally, waiting just a few moments for the light to change will improve a photograph. On a partly cloudy day, for example, lighting will alternate between diffuse and directional. One can have diffuse light by waiting for clouds to pass overhead or directional sunlight by waiting for clouds to clear. Starting a few hours earlier or later so that sunlight comes from a different angle can also improve one's control over outdoor-lighting effects. Although this may not be a reliable nor efficient procedure, many truly great outdoor photographs could only have been made in this way and would not have been possible without great patience, diligence, and foresight on the part of the photographer.

Lighting by Daylight

As discussed earlier, intense light sources simplify high-resolution photography by allowing the photographer to use optimum camera controls, fast shutter speeds to reduce blurring from camera or subject motion, and small apertures for extensive depth of field, while using the slower fine-grain films for maximum resolving power. Nothing in photography will improve image quality more than good illumination will.

Increasing the Light Level

Electronic flash units have three qualities that make them especially valuable in high-resolution work: (1) their light can be very intense; (2) they are nearly point sources; and (3) they can be designed to operate at nearly instantaneous speeds. Flash durations of less than a billionth of a second have been achieved using electronic flash

High-Speed Flash

units. The duration of the flash is determined by the capacitance and resistance in the electronic circuitry of the flash unit:

$$t = \frac{1}{2}RC$$

where t is flash duration in seconds, R is resistance in ohms, and C is capacitance in farads. Electronic flash units are especially well suited for combatting blurring from subject motion.

High-speed flash units are easy to come by if they do not need a very high light output. Some of the low-powered, inexpensive units designed for general photography have flash durations as short as 1/50,000th second (20 microseconds). Because of their low light output, however, they are useful only when the subject is no farther than a few feet from the camera. They are made to order for close-up photography where subject distances range from a few inches to a few feet. (See Plate 13.) In such applications they allow use of slow, fine-grain films at aperture settings as small as f/16 or f/22.

High-wattage professional flash units provide greater light output than small amateur units, but they usually operate at speeds closer to 1/500th second unless they have been specially designed for high-speed output. Obtaining high lighting output as well as short flash duration calls for elaborate circuitry featuring low capacitance and high voltage. A high-speed, high-output unit may use condensers with capacitance ratings of one or two microfarads and power ratings of five to ten kilovolts. Light output from such a unit is typically about 50 joules at a flash duration of about one microsecond. The condensers in such a flash unit are bulky, heavy, and expensive. They also store a potentially lethal electric charge. High-speed, high-output flash units must be used with care and with some knowledge of the hazards associated with high voltages and heavy current flow.

Flashbulbs, particularly the large M3 bulbs and the powerful Press 50 bulbs, provide very good illumination and do so with less bulk and weight than do electronic flash units of comparable output. In dim lighting, bulbs may offer a feasible solution to obtaining high-quality images. A General Electric number 50 bulb, for example, is no larger than an ordinary electric lightbulb, but it can be fired using flashlight batteries. It has a guide number of 1000 when used with an ISO 400-speed film. This means that, used at an aperture of f/1.4, one bulb is bright enough to light a subject on the far side of a football stadium—from the long end.

When high-output flashbulbs are used at closer range, a slow, fine-grain film or a smaller aperture can be used. Even in difficult situations, flashbulbs can effect a great improvement in resolving power. However, because of the increasing popularity of electronic flash units, large flashbulbs like the number 50 and its associated fixtures are not in great demand anymore and are hard to find.

Incandescent lamps offer another way to increase the level of available light. Their disadvantage is that they generate heat and they require special power or field generators for location photography. Their advantage is that they can be used in long exposures, allowing more flexibility in choosing an aperture setting.

Chapter 22

Ground Resolution
Resolving Specific Subject Details

In surveillance photography, investigative photography, and the like, in which conditions may be less than ideal for making high-resolution photographs, one's objective is often to achieve a recognizable likeness of a human subject or to be certain that specific image features of known size will be clearly resolved in a photograph. The assignment may be to photograph an airplane so that its rivets can be counted, or copy a document so that each character can be deciphered, or photograph a person at long range as evidence in an investigation. The concept of ground resolution allows one to evaluate these assignments in advance to determine how good the camera system must be for success.

Whenever the size of the feature to be resolved is known or can be estimated, ground resolution can be used to determine in advance whether a satisfactory record can be made of that feature. If you have already made an unsatisfactory photograph, you can determine the conditions that must be changed to succeed with the assignment, or you can determine whether your assignment is theoretically impossible.

The Ground-Resolution Equation

Ground resolution is a term originally used in aerial photography to refer to the size of ground objects resolved in an aerial photograph. In the present context it has a more general meaning. It refers to the smallest surface feature, or to the smallest separation between subject features, resolved in a photograph. See Figure 22.1. The concept is derived from one introduced by Carl M. Franz of the United States Naval Surface Weapons Center.

The value on the left side of the equation, g, represents ground resolution or subject-detail separation; it represents the size of the smallest subject feature that can appear in a photograph fully resolved. The spread function S represents the size of the corresponding image point as determined by the level of degradation in the system. The relationship between g and S is influenced by the focal length of the lens F and by the distance u from the subject to the camera as follows:

$$g = \frac{uS}{F}$$

Using this formula you can determine the conditions that must prevail if you are to photograph successfully the rivets in the aircraft, the document, or the person at a distance. You can, that is, if the value of g, the size of the subject feature to be resolved, is known. In general, g must be set to the thickness of the smallest subject feature to be resolved. The smaller its value the clearer the image. To resolve the rivets, assuming they are about a quarter of an inch in diameter, g must evaluate to one-quarter inch or a little less. To resolve the person's facial features, g must be smaller than the dimension of the smallest facial feature to be recorded. If the photograph must contain details as small as an eyelash, g must be equal to or smaller than the thickness of a single eyelash.

Whenever g can have a greater value, perhaps the size of an eyebrow instead of an eyelash, you will be able to use a system of lower resolving power—a faster film, for example, or a larger aperture—or you will be able to increase the working distance between the camera and the subject.

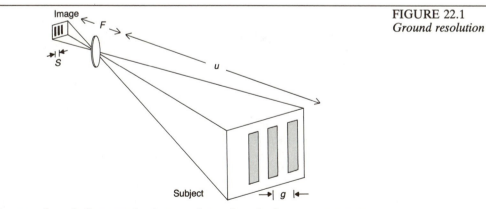

FIGURE 22.1
Ground resolution

The ground-resolution equation lets one determine whether a camera system of known resolving power can resolve details of a given size at a given distance. The smallest subject feature g that can be successfully rendered in a photograph is determined by camera-to-subject distance u, the focal length of the lens F, and the camera system's spread radius S.

It is clear from this equation that the simplest way to resolve small subject features or to reduce the value of *g* is to reduce the size of *u*, camera-to-subject distance. In other words, move in closer to the subject. If a subject can be photographed from a close enough distance, it is possible in theory to record any of its features, down to microscopic size or down to the resolution limit of the optical system.

When you must work from afar, there will be a limit on the level of detail you can record. For example, if you have a 50mm lens that resolves 85 lines per millimeter and you must capture the level of detail visible to the unaided eye (70 microns), you must be within a foot or so of the subject. If you must be farther away, the ground-resolution formula lets you determine the greatest distance at which you can successfully record specific subject details. If this distance is closer than you can safely come, the equation helps you to select the lens focal length or predict the resolving power that makes the camera system adequate.

In aerial photography, military surveillance, industrial intelligence, and many scientific applications of photography, the photographer is often unable to move in on the subject. It is the nature of such applications that approaching a subject too closely may endanger the success of the assignment or may endanger the safety of the photographer. Often the photographer must avoid detection and may be physically restrained by fences or barriers. Furthermore, there may be only one opportunity to make the photograph. On such an assignment, it is imperative that you know in advance whether your system can resolve features of interest at the required distance.

Applying Ground Resolution

To apply the ground-resolution equation you must establish the values of three of the parameters in the formula and compute the fourth. In a typical situation, you might begin by determining the size of the smallest subject feature to be recorded. Table 22.1 and Figure 22.2 suggest values to use when alphanumeric characters and human subjects are involved. Next you must know the spread function of the photographic system. If you have not already done so, measure the resolving power of the system you will use and compute its spread radius as shown in Chapter 8. Finally, substitute these values into the equation to determine the optimum camera-to-subject distance or the optimum lens focal length needed to obtain a given level of detail. These steps will now be examined in detail.

Select Subject-Detail Separation

When you work with unfamiliar subjects, determining an appropriate value for subject-detail separation will call for judgment and perhaps trial and error. You can see from Table 22.1, for example, that a

Objective	Minimum subject-detail separation in centimeters	
Detection: To determine whether people are present	3–15*	TABLE 22.1 *Subject-detail separation for human subjects*
Recognition: To recognize familiar faces, distinguish features	1.0	
Positive identification: To identify a face with confidence	0.8	
Judicial identification: To identify a person with judicial sufficiency	0.4	

*A human figure can be detected at a subject-detail separation of 15cm only when it contrasts clearly with the background. Detection will be easier when the smaller value is resolved. At a detail separation of 3cm, people are detected as people even under conditions of poor contrast and heavy background clutter. At 15cm, a human figure appears as nothing more than an elongated spot and may be assumed to be a person only if on a sidewalk or where people are likely to be.

subject-detail separation of between 3 and 15cm is adequate for detecting the presence of human subjects. The far end of this range, 15cm, is adequate only when you know or have reason to believe beforehand that human subjects are present at the location photographed, and you must simply determine how many.

If ground resolution is no better than 15cm (about 6 inches), subject features smaller than 15cm will not be visible in the photograph. The human torso will reproduce above the resolution threshold and should be visible, but none of its identifying features will be captured. Arms and legs, for example, are generally smaller than 15cm across and will not be resolved. A human form will be barely

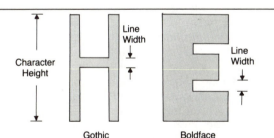

FIGURE 22.2
Alphanumeric subject-detail separation

Gothic   Boldface

The separation needed to resolve alphanumeric characters can be determined from the thickness of the lines or from the height of the characters. If the characters are boldface, measure the smallest width to be resolved, be it a line or space, or use one-seventh the height of a character. This much separation guarantees sufficient legibility to read words and to distinguish all but the difficult letters, like *G*. To distinguish any letter, set subject-detail separation to 0.8 times line thickness or to one-tenth the height of a character. If you must determine only that writing is present, you may set separation to 1.4 times line width or to one-fifth the height of a character. These values are minimums. If the system resolves smaller details, so much the better.

distinguishable from a post or other dissimilar object. It may not be possible to determine from the image alone whether or not the figure is really that of a person.

Identifying a likeness as a specific person requires that a system record much finer detail. To render faces recognizable a system must resolve subject details of about 1cm or 1/2 inch. To render faces recognizable for judicial purposes, good enough for positive identification, resolution must be at least 0.4cm. Mind you, accepting subject-detail separations this large will not yield portrait-quality images; only very coarse features are resolved at this level of detail.

Determine the Spread Radius

Before the equation for ground resolution is applied, the composite-spread function or spread radius of the camera system you use must be known. If you own several lenses and plan to use them on assignment, separate resolving-power tests should be made and spread radii computed for each likely combination of lens and film. This is important—switching to a different lens or film can change the value of the spread function and alter the computation.

Of the three approaches suggested by the equation for improving the ground resolution of a system, the most useful is to improve its resolving power or reduce its spread function. That is, use a fine-grain film, use a tripod, use the lens at its optimum aperture setting, or apply some other high-resolution technique.

Select the Optimum Lens Focal Length

As noted earlier, details of nearly any size can be resolved in a photograph if the camera can be placed close enough to the subject. This is true even with cameras or lenses of low quality. When getting close is impractical or unsafe, an alternative is to use a long focal-length lens. Telephoto lenses improve the resolution of detail in one respect by giving an enlarged image of subject features, but in other respects they reduce resolution. As shown in Chapter 19, high-powered telephoto lenses are difficult to use. They magnify blurring, making it difficult to compensate for camera motion, and they give shallow depth of field. They also tend to have lower inherent resolving power than lenses of standard focal length. Experience shows that little is gained by using a longer focal length than needed to achieve the required ground resolution. Nevertheless, there are difficult situations where a long focal-length lens is the only way to achieve adequate subject detail. Use these lenses when you must, but be careful with them.

If a wide area must be included in the frame, a strong telephoto lens may too severely restrict the field of view. Given no other choice, you can use a long focal-length lens and step back from the subject far enough to get the coverage you need, but this usually yields less image detail than selecting a lens of shorter focal length and moving in close. The procedure likely to yield the greatest clarity is to use a medium focal-length, high-resolution lens at a distance that provides the required subject coverage.

Having the capability to predict ground resolution becomes invaluable to a photographer who uses his or her system at the limits of its resolution potential and information must be recorded at any cost, even if it is barely decipherable. There are times when raw information content is more important than perfect definition and elaborate detail—when, as long as a message can be read or an image deciphered, sharpness does not matter. On these occasions marginal resolution is certainly preferred over losing the image entirely.

For example, consider a case where a photographer must record the name on the hull of a boat at sea but cannot leave the shore. From an estimate of the size of the characters or the thickness of the lines to be resolved and from a knowledge of the resolving power of the camera system used, the photographer can determine whether a normal focal-length lens can record the image successfully, whether a longer focal-length lens will be needed, or whether the task will be impossible from that distance.

Marginal Detail Separation

Sometimes more than one requirement constrains the photographer. Consider a case in which a stake-out camera is set up to show vehicles arriving at a building. The objective of the photograph is to show the building or enough of it to prove which building it is. At the same time, the license plate numbers on arriving vehicles must be resolved. The need to show the building ties the photographer to a fixed combination of camera-to-subject distance and lens focal length. The only workable option using a given focal-length lens will be to place the camera at a distance that fills the negative with the image of the building. If the subject-detail separation computed at this distance is inadequate to resolve the characters on the vehicle license plate, a system with greater resolving power must be used because the objective of the project cannot be met using the original system.

The uses made of the ground-resolution formula need not be exotic. Consider a situation in which you must photograph a group of 800 people so that each can be recognized in an enlarged print. This assignment requires that subject details larger than about 4–8mm be resolved. How can the success of the assignment be guaranteed?

In this situation you face a problem similar to that in the stake-out. The lens must have wide enough coverage or the camera must be far enough back to take in the entire group. You can choose the focal length of the lens or the camera-to-subject distance, but having selected one parameter, you have no real control over the other. To succeed at the task you must be certain that a system has been selected whose resolving power is adequate. If your professional reputation rests on the success of the shot, you will want to confirm that it is indeed adequate by computing ground resolution.

Dual Constraints

Chapter 23

Afterword
Image Clarity in Brief

This afterword summarizes the key ideas presented and shows how they point toward a reasoned approach to high-resolution technique.

Image Degradation

An image will never represent its subject more accurately at a particular stage of image formation than it did at any earlier stage. Indeed, if the photographer does not act specifically to counter it— in every stage of the imaging process—a decline in image information is inevitable.

The dimensions of an isolated point-sized image turns out to be a useful and valid measure of a decline in image quality. If an imaging system were perfect, the image points it creates would be infinitely small, without measurable size. Measurable size of any amount is thus a direct indication of how much an image has been degraded.

Each item of equipment used in an image-forming process and each operation on an image, regardless of the quality of the component or the care with which the operation is performed, has a degrading influence that causes the image of an infinitesimal point to reproduce as a spot of finite size. The mathematical descriptions of these degrading components are known as spread functions.

The way various spread functions combine to degrade the imaging quality of a photographic system is predicted in the *degradation formula*:

$$S^2 = s_o{}^2 + s_e{}^2 + \ldots + s_x{}^2$$

This formula shows that degradation is the active agent in image formation and it suggests a workable approach to the practice of high-resolution photography: making sharp, detailed photographs involves not bringing quality to an image but reducing influences that degrade quality. Put another way, the high-resolution photographer must not assume that he or she is to make a degraded image better. One should understand that the image arriving at the lens is as faithful to the subject as it will ever be. The task then becomes to minimize deterioration of that image. To succeed, one must understand how degradation operates and must know how to control each of its sources.

The degradation formula demonstrates the need for the *concomitant reduction of degradation,* which means that all sources of degradation must be identified and eliminated from an imaging system before the system can reach its peak resolution potential. Photographers who limit their scope and focus their attention on obvious kinds of degradation, and nothing more, can hardly expect to achieve high-resolution results. They must attend to them all. Any kind of degradation unchecked, originating from any source, can alone destroy image clarity. One's understanding of the mechanisms of image degradation must be complete and one's control over them must be concurrent.

Failing this, the high-resolution photographer must work to reduce the largest spread component before acting on lesser ones. The presence of one or more large spread components imposes a resolution limit on the imaging system. As long as this limit is active, it will neutralize refinements brought about by improvements to smaller spread components.

The efforts of a beginning photographer will typically be constrained by resolution limits in a predictable order of progression: image displacement, emulsion turbidity, and optical spreading. The photographer will gain most and improve skills fastest by acting on these resolution limits in the order listed. In other words, one must eliminate the massive damage done by image motion and vibrations in order to appreciate the more moderate damage done by fast, coarse-grain films. One must use a high-resolution, fine-grain film to detect the subtle damage done by the aberrations of today's fine lenses. One must use the finest optics to appreciate the damage done by diffraction and other subtle kinds of degradation.

Equipment Quality

Although photographers can do nothing to reverse existing degradation, they will usually be able to make better photographs by upgrading their imaging system and by selecting the best components available. The lens and film are especially critical components. A solid tripod and a good lens hood are also valuable accessories. Caution is needed in selecting a system intended for high-resolution photography. Photographers who allow themselves to be distracted by the quest for speed, faster lenses and faster films for their own sake, can easily compromise their system's imaging quality.

Also, limits are imposed by technology and by physical laws that determine how far improvements in equipment can be taken. A system good enough to be sensitive to the effects of diffraction is about as good a system as a high-resolution photographer can hope to have. Starting with such a system, the photographer should be concerned not about equipment, but about technique.

Photographic Craftsman-ship

Photographers must be certain that all the detail and sharpness they need in a photograph has been captured in the negative. Information that is lost at the camera stage is lost forever. Disregarding this principle, many photographers overrate the effectiveness of film development and print processing, somehow believing that they can use these processing steps to compensate for inept camera techniques. Film development can be tailored to modify the density range of a negative, matching it to the exposure range of a particular print paper. Tonal adjustments can be made at the printing stage to expand the tonal scale of a print. Short of recording a new image, however, image detail cannot be improved upon after an image has been recorded on film.

Good camera technique is essential. It involves the correct use of aperture and shutter speed to arrive at precise exposure, adequate depth of field, and reasonable control over image motion. It involves accurate focusing and skillful image management.

One must also consider how contrast influences the resolution of detail. In exploring the mechanics of image-point spreading within the microstructure of a photograph, resolution was found to depend indisputably on contrast. Consequently, effective lighting, correct exposure, and accurate tone control are more necessary to high-resolution technique than has generally been supposed.

Factors favorable to clarity are bright lighting, medium or short focal-length lenses, full-frame imagery, good subject contrast, and steadiness or freedom from motion at the camera and subject. Such a situation allows use of a slow, fine-grain film and an optimum lens aperture, making it possible to create the best images the silver-halide system can produce. Under less than ideal circumstances, special resolution limits will be encountered, and one must act at once to isolate and control the limiting spread component.

Having learned to control the common sources of image degradation, you may ultimately reach the level of craftsmanship where diffraction is the final factor that limits the resolving power of your system. If so, your images will be about as sharp and detailed as they can conceivably be. You will then be in a position to appreciate more fully the finer points of high-resolution technique. You will know to select your aperture not only in consideration for exposure, but for overall image sharpness. You will understand the value of a rigid, stable camera support. You will be aware of light and its influence on the resolution of detail. You will guard against flare and spurious light that degrades contrast. Every aspect of your photographic technique will reflect your new awareness.

Achieving success in high-resolution photography requires commitment, discipline, and foresight. The following hypothetical example demonstrates the difficulties and shows the need for a firm personal commitment by those who would achieve clarity where clarity is the paramount consideration.

Diminishing Returns

Begin by considering the performance of a system capable under ideal conditions of resolving 100 lines per millimeter. In general photography this is a system of superb quality. In the hands of a beginner with little or no training or experience, however, it will likely resolve no more than 5 to 10 lines per millimeter. A typical beginner will hand hold the camera and do so poorly. Handling of the camera will be awkward and jerky; camera movement will be severe. After learning to grip the camera and operate its controls smoothly, the photographer may achieve 20 lines per millimeter and may reach 40 lines per millimeter after developing a flawless hand-holding technique and learning to exploit faster shutter speeds.

Then, using a tripod, the photographer may reach 80 lines per millimeter routinely, though many photographs will resolve less. At this point camera movement will still account for failing to reach the full potential of the system, but it will be just one cause.

In spite of using a rock-solid tripod, several things can prevent the photographer from attaining the highest level of quality. For example, the use of too large or too small an aperture may bring lens faults or diffraction into play. Slight errors in focusing may arise. Microvibrations, common in single-lens reflex cameras, may occur. Or if exposure time is too long, ground vibrations may become a factor. In none of these situations will a tripod alone prevent degradation from occurring. By correctly using massive and stable camera supports, by selecting a film to suit the assignment, by setting the aperture and shutter speed with deliberation, and by correctly attending to depth of field and focusing, the photographer may at times resolve as much as 95 lines per millimeter on field assignments.

Then, working in studio conditions in which all variables in the process are within control—where the environment is calm, ground vibrations have been minimized, and lighting is good—the photographer may reach 98 lines per millimeter. With luck and with nothing going wrong, he or she may reach 99 lines per millimeter, though this is difficult to do consistently. Even the peak resolving power of 100 lines per millimeter may be attained—or may not be.

The numbers are hypothetical. They do, however, reflect the operation in photography of the law of diminishing returns. By being conscious of each hazard that reduces image quality and combatting them all, one can master high-resolution photography. Success, however, is obtained only in degrees of geometrically increasing difficulty. When resolving power is near its theoretical maximum, achieving marginal increments in quality requires a disproportionate expenditure of time, money, and effort. Put another way, a fixed expenditure will not likely yield the same increase in quality in each successive step. The very last increment, the one that makes the image as good as it can be, will be especially difficult and costly to attain.

This is why the strength of one's commitment to high-resolution technique is so important. Mastering all the subtleties of high-resolution photography requires a single-minded dedication to image quality over matters of convenience and to resolving power over other categories of performance. Obtaining high-resolution images requires a commitment that transcends the desire for "good" photographs. Photographers do not burden themselves with sandbags, lead weights, and massive tripods to get a photograph that is merely good. Good images can be created without them. The commitment is like a compulsion to permit no obstacle to hinder your quest for immaculate images. You can improve your skills without such a commitment, but you cannot sustain a record of technical excellence.

Bibliography

Adams, Ansel. *The Negative: Exposure and Development*. New York: Morgan & Morgan, 1948.

————. *The Print: Contact Printing and Enlarging*. New York: Morgan & Morgan, 1968.

Blaker, Alfred A. *Field Photography*. San Francisco: W. H. Freeman, 1976.

Bomback, Edward. *Manual of Photographic Lighting*. London: Fountain Press, Ltd., 1971.

Burden, James W. *Graphic Reproduction Photography*. New York: Hasting House, 1940.

Butts, Allison, and Coxe, C. D. *Silver, Economics, Metallurgy and Use*. New York: Van Nostrand Reinhold, 1967.

Carrol, John S. *Photographic Lab Handbook*. New York: Morgan & Morgan, 1974.

Carroll, Higgins, and James. *Introduction to Photographic Theory*. New York: Wiley, 1980.

Conrady, A. E. *Applied Optics and Optical Design*. Mineola, N.Y.: Dover, 1957.

Cox, Arthur. *Optics, The Technique of Definition*. Boston: Focal Press, 1966.

Dake, and DeMent. *Fluorescent Light and Its Applications*. Brooklyn, N.Y.: Chemical Publishing Co., 1941.

Dalladay, A. J., ed. *The British Journal of Photography Annual, 1967*. London: H. Greenwood & Co, 1967.

Dalton, Stephan. *Caught in Motion: High-Speed Nature Photography*. New York: Van Nostrand Reinhold, 1984.

Davis, Phil. *Beyond the Zone System*. New York: Van Nostrand Reinhold, 1981.

Eastman Kodak Company. *Clinical Photography*. Rochester, N.Y.: Kodak, 1972.

————. *Using Photography to Preserve Evidence*. Rochester, N.Y.: Kodak, 1976.

————. *Fire and Arson Photography*. Rochester, N.Y.: Kodak, 1977.

————. *Processing Chemicals and Formulas*. Rochester, N.Y.: Kodak, 1977.

————. *Basic Sensitometry Workbook*. Rochester, N.Y.: Kodak, 1981.

————. *High-Speed Photography*. Rochester, N.Y.: Kodak, 1981.

———. *Kodak Films—Color and Black-and-White*. Rochester, N.Y.: Kodak, 1981.

Eaton, George. *Photographic Chemistry*. New York: Morgan & Morgan, 1981.

Elliot, and Dickson. *Laboratory Instruments*. New York: Chemical Labs Publishing Co., 1960.

The Focal Encyclopedia of Photography. London: Focal Press, 1969.

Franz, Carl M. *The Photographer's Guide*. Silver Springs, MD.: Naval Surface Weapons Center, Photographic Engineering and Services Branch, 1976.

Habell, and Cox. *Engineering Optics*. Pittman & Sons, 1948.

Handbook of Optics. Optical Society of America, 1978.

Hawkins, R. *Production of Microfilms*, vol. 5, pt. 1, "State of the Library Art."

Hedgecoe, John. *Advanced Photography*. New York: Simon & Schuster, 1982.

———. *The Photographer's Handbook*. New York: Alfred A. Knopf, 1978.

Holtz, John. *In Focus*. New York: Harper and Row, 1980.

Jacobson, K., and Manheim, L. A. *Enlarging*. London: Focal Press, 1967.

James, T. H. *Theory of Photographic Process*. New York: Macmillan, 1977.

Kettlkamp, Larry. *Tricks of Eye and Mind*. New York: William Morrow, 1974.

Langford, Michael J. *Advanced Photography*. London: Focal Press, 1974.

———. *Basic Photography*. Boston: Focal Press, 1973.

Lyalikov. *The Chemistry of Photographic Mechanisms*. London: Focal Press, 1960.

Mason, L. F. A. *Photographic Processing Chemistry*. London: Focal Press, 1975.

Mees, C. E. K. *From Dry Plates to Ektachrome: A Story of Photographic Research*. New York: Ziff-Davis, 1961.

Neblette, Carrol B. *Fundamentals of Photography*. New York: Van Nostrand Reinhold, 1970.

——— and Murray, Allen E. *Photographic Lenses*. New York: Morgan & Morgan, 1973.

Patterson, Freeman. *Photography and the Art of Seeing*. New York: Van Nostrand Reinhold, 1979.

Photo Lab Index. New York: Morgan & Morgan, updated continuously.

Picker, Fred. *The Zone VI Workshop*. New York: Amphoto, 1978.

Rayleigh, J. W. S. *Collected Papers of Lord Rayleigh*. Cambridge, Eng.: Cambridge University Press, 1902.

Sansone, Sam J. *Police Photography*. Cincinnati, OH: Anderson Publishing, 1977.

Schaefer, John P. *How to Use the Zone System for Fine Black-and-White Prints*. H.P. Books, 1983.

Siljander, Raymond P. *Applied Surveillance Photography*. Springfield, IL.: Charles C Thomas, 1975.

Skudrzyk, Eugen J. *Photography for the Serious Amateur*. New Jersey: A. S. Barnes, 1971.

SPSE Handbook of Photographic Science and Engineering. New York: Wiley, 1973.

Stevens, G. W. W. *Microphotography*. London: Chapman & Hall, 1968.

Sturge, John M., ed. *Neblette's Handbook of Photography and Reprography Materials, Processes and Systems*. New York: Van Nostrand Reinhold, 1977.

Swedlund, Charles. *Photography: A Handbook of History, Material, and Processes*. New York: Holt, Rinehart & Winston, 1974.

Time-Life Books. *The Camera*. New York: Time Inc., 1970.

————. *Color*. New York: Time Inc., 1970.

————. *Light and Film*. New York: Time Inc., 1970.

————. *Photography as a Tool*. New York: Time Inc, 1970.

————. *The Print*. New York: Time Inc., 1970.

Todd, Hollis, and Zakia. *Photographic Sensitometry*. New York: Morgan & Morgan, 1981.

Vestal, David. *The Craft of Photography*. New York: Harper & Row, 1978.

White, Minor, and Zakia, Richard. *The New Zone System Manual*. New York: Morgan & Morgan, 1984.

Whittaker, Leathem. *The Theory of Optical Instruments*. New York: Hafner.

Index